Galina Filipuk, Andrzej Kozłowski
Analysis with Mathematica®

Also of Interest

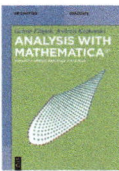

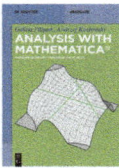

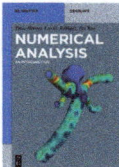

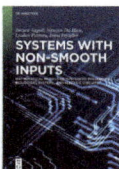

Galina Filipuk, Andrzej Kozłowski

Analysis with Mathematica®

Volume 3: Differential Geometry, Differential Equations, and Special Functions

DE GRUYTER

Mathematics Subject Classification 2010
53A04, 53A05, 97I40, 34A05, 34M03, 34M55, 33C75, 33C47, 33C99, 97I80, 97N80

Authors

Dr. hab. Galina Filipuk
University of Warsaw
Faculty of Mathematics, Informatics and
Mechanics
Banacha 2
02-097 Warsaw
Poland
Filipuk@mimuw.edu.pl

Dr. Andrzej Kozłowski
University of Warsaw
Faculty of Mathematics, Informatics and
Mechanics
Banacha 2
02-097 Warsaw
Poland
akoz@mimuw.edu.pl

ISBN 978-3-11-077454-2
e-ISBN (PDF) 978-3-11-077464-1
e-ISBN (EPUB) 978-3-11-077475-7

Library of Congress Control Number: 2022930269

Bibliographic information published by the Deutsche Nationalbibliothek
The Deutsche Nationalbibliothek lists this publication in the Deutsche Nationalbibliografie; detailed bibliographic data are available on the Internet at http://dnb.dnb.de.

© 2022 Walter de Gruyter GmbH, Berlin/Boston
Cover image: Created by the authors with the help of Mathematica®
Typesetting: VTeX UAB, Lithuania
Printing and binding: CPI books GmbH, Leck

www.degruyter.com

Preface

This volume is devoted to applications: differential geometry of curves and surfaces with remarks about visualization in Mathematica®,[1] differential equations, and elements of special functions. Functions of one and more variables were considered in Volume 1 [7] and Volume 2 [8]. We continue to demonstrate that Mathematica® can be used effectively as an aid in solving mathematical problems or at least in discovering their solution. As before, we leave most proofs and many other details to several recommended textbooks and various internet resources, and we also assume that the reader knows the basics of the Wolfram Programming Language™. We refer to the preface of Volume 1 for more information.

This book is partially based on three courses at the University of Warsaw (the undergraduate course *Curves and Surfaces with Mathematica®* and the graduate courses *Introduction to elliptic functions* and *Ordinary differential equations: methods and applications*). In this book we explain in detail a new feature of Mathematica®, the built-in database Entity on examples of plane and space curves and surfaces. We use and explain a lot of new recently built-in functions, for instance, Region, and so on. To give a lot of insight into differential geometry, we extensively make use of Manipulate to create informative interactive graphics.

When writing Volumes 1 and 2 we used Mathematica®'s versions 11 and 12, whereas this volume is based on version 12.3 which was released while writing this book. In this volume we often use InputForm and StandardForm for both inputs and outputs.

We hope that this volume will inspire the reader to experiment with other topics. This book is primarily intended for educational purposes, but we believe that some parts of it can be of interest to researchers.

The citation of registered names, trade names, trade marks, etc., in this work does not imply, even in the absence of a specific statement, that such names are exempt from laws and regulations protecting trade marks, etc., and therefore free for general use.

The various Wolfram trademarks and screenshots are used with the permission of Wolfram Research, Inc., and Wolfram Group, LLC.

---

[1] http://www.wolfram.com/

https://doi.org/10.1515/9783110774641-201

Contents

Preface —— V

1   Graphics in Mathematica® —— 1
1.1   Built-in functions —— 1
1.2   Example 1: Roots of unity —— 38
1.3   Example 2: The complex exponential map —— 41
1.4   Example 3: Creation of a torus by rotating a circle —— 46

2   Regions in Mathematica® —— 49

3   Differential equations —— 71
3.1   Differentiation in Mathematica® revisited —— 71
3.2   Built-in functions to solve differential equations and systems —— 73

4   Differential geometry of curves and surfaces in Mathematica® —— 89
4.1   The Wolfram Knowledgebase —— 89
4.2   Vectors in Mathematica® —— 99
4.3   Curves —— 101
4.3.1   A physical interpretation of a curve in $\mathbb{R}^n$ —— 101
4.3.2   Implicitly defined curves —— 107
4.3.3   Parametric plots of curves in 2D and 3D —— 111
4.3.4   Length of parametric curves and the function ArcLength —— 117
4.3.5   Unit (arclength) parametrization —— 119
4.3.6   Curvature of a curve in $\mathbb{R}^n$ —— 124
4.3.7   Signed curvature of a plane curve —— 127
4.3.8   Evolutes and involutes for plane curves —— 129
4.3.9   Curvature of implicitly defined plane curves —— 137
4.3.10   Drawing plane curves with assigned curvature —— 146
4.3.11   Envelopes —— 147
4.3.12   Curves in $\mathbb{R}^3$ —— 152
4.4   Surfaces —— 161
4.4.1   Drawing surfaces using curves —— 162
4.4.2   A tangent space and a normal to a smooth surface —— 170
4.4.3   Metrics on a surface: The first and second fundamental forms —— 172
4.4.4   Curves and frames on a torus —— 183
4.4.5   Geodesics —— 187
4.4.6   The tangent cone —— 191

5   Elements of the theory of special functions —— 197
5.1   Linear special functions —— 197

5.2   The Gauss hypergeometric function —— **201**
5.3   Legendre polynomials —— **203**
5.4   Example: Discrete orthogonal polynomials —— **206**
5.5   Nonlinear special functions and Painlevé transcendents —— **208**

6   **Elliptic functions —— 215**
6.1   The generalized sine function —— **215**
6.2   Algebraic addition theorems —— **229**
6.3   The Weierstrass elliptic function —— **231**
6.4   Example: Addition formulas for the Weierstrass elliptic function —— **239**

7   **Elements of complex analysis —— 241**
7.1   Complex numbers and functions —— **241**
7.2   Example 1: Möbius transformation —— **243**
7.3   Example 2: The conformal mapping $w = z^2$ —— **243**
7.4   Example 3: The Cauchy residue theorem —— **260**

Bibliography —— **263**

Index —— **265**

1 Graphics in Mathematica®

When we study differential geometry or differential equations, it is useful to be able to draw curves or surfaces in the plane or in space, or solution curves and vector fields. Mathematica® has a lot of built-in functions for this purpose, with a variety of options. We shall not present all of them here, but highlight a couple of useful ones which we will use later on. Moreover, we will explain various ways in which objects can be colored. Visualization is of particular importance if one wants to create dynamic illustrations similar to those appearing at Wolfram Demonstrations Project.[1]

1.1 Built-in functions

The most fundamental built-in function is Plot or Plot3D. It has a lot of options (since the list is very long, we shall not present the output). Each option is described in the documentation center in detail.

In[·]:= {Options[Plot] // Length, Options[Plot3D]// Length}
Out[·]:= {63, 82}

For instance, we can choose PlotRange for our graph and compare the following plots of the same function

In[·]:= Grid[{{Plot[x^2, {x, -1, 1}], Plot[x^2,
        {x, -1, 1}, PlotRange -> {0, 5}]}}]

Out[ ]=

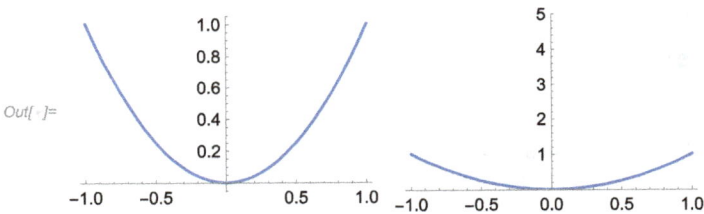

Figure 1.1

We can also fill the region under the graph of the function

In[·]:= Plot[Sin[x], {x, 0, 1}, Filling -> Bottom]

---

1 https://demonstrations.wolfram.com.

https://doi.org/10.1515/9783110774641-001

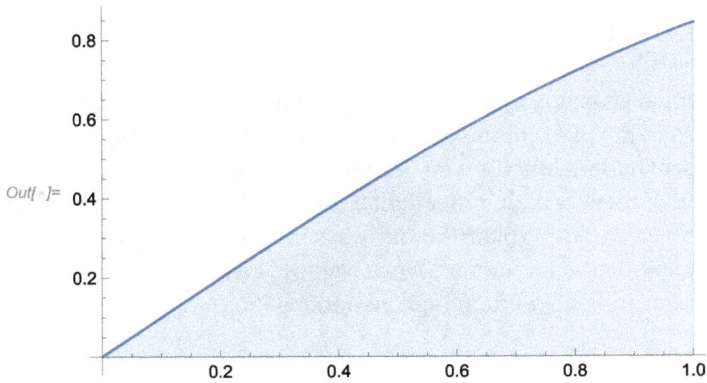

Figure 1.2

The reader should try experimenting with various options. Here we give only a few examples.

In[·]:= Plot[Sin[x], {x, -1, 1}, ColorFunction ->
          "Rainbow"]

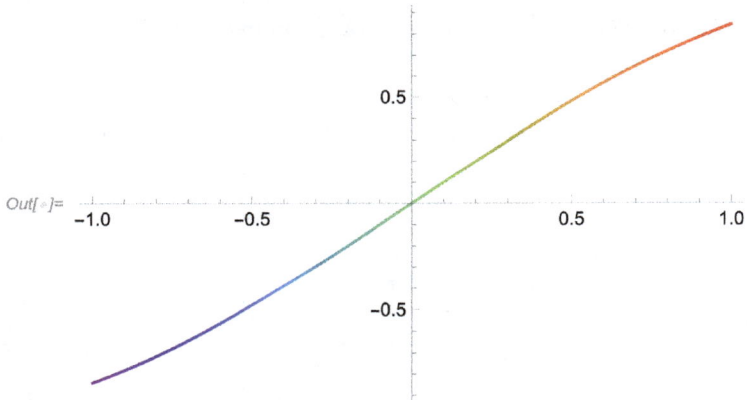

Figure 1.3

In[·]:= Plot[Sin[x], {x, -1, 1}, PlotStyle ->
          {Opacity[0.3], Green}]

Out[ ]=

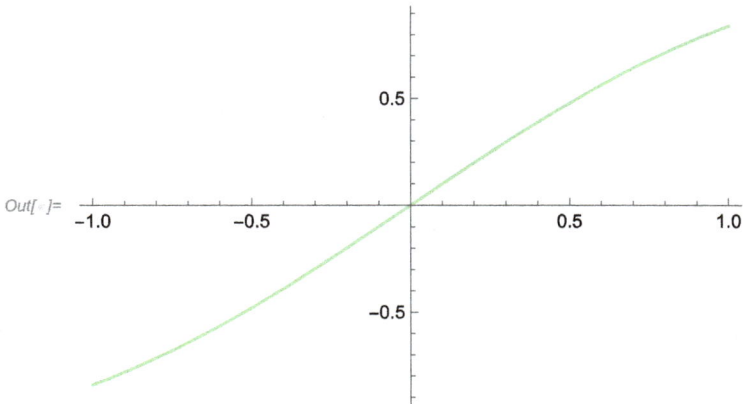

Figure 1.4

The built-in `ParametricPlot` and `ParametricPlot3D` allow us to draw parametric curves on plane or in space:

In[ ]:= `ParametricPlot[{t, Sqrt[t]}, {t, 0, 3}]`

Out[ ]=

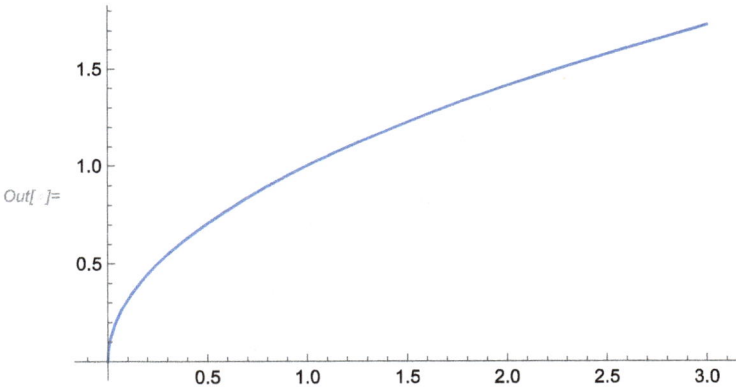

Figure 1.5

In[ ]:= `ParametricPlot3D[{t, Sqrt[t], t^2}, {t, 0, 3}]`

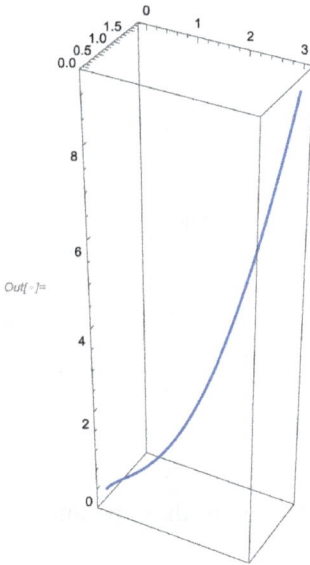

Figure 1.6

Curves given by implicit equations can be drawn using ContourPlot:

```
In[·]:= ContourPlot[x^2 + y^2 == 1, {x, -1, 1},
        {y, -1, 1}, PlotRange -> Automatic]
```

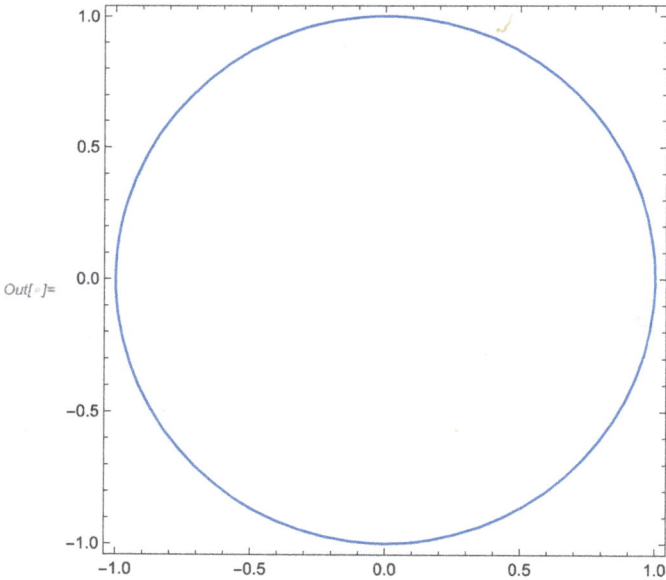

Figure 1.7

We can also combine different curves in one picture. Note that we have chosen different intervals over which to plot the parabola and the circle, and that Mathematica®
drew both over the smaller interval:

```
In[·]:= Show[Plot[x^2, {x, 0, 1}], ContourPlot[
        x^2 + y^2 == 1, {x, -1, 1}, {y, -1, 1},
        PlotRange -> Automatic]]
```

Out[·]=

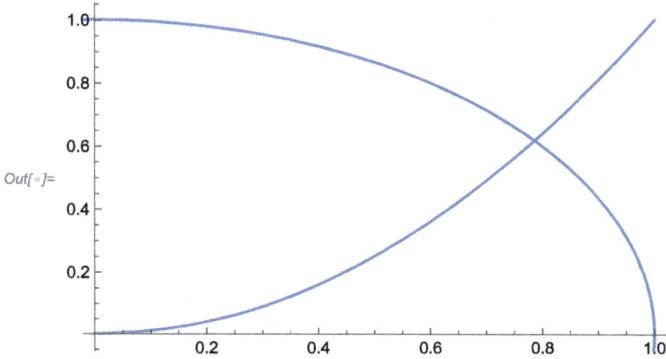

Figure 1.8

There are a lot of built-in graphics primitives like Circle discussed below in more
details:

```
In[·]:= Graphics[Circle[{1, 0}, 1], Axes -> Automatic]
```

Out[·]=

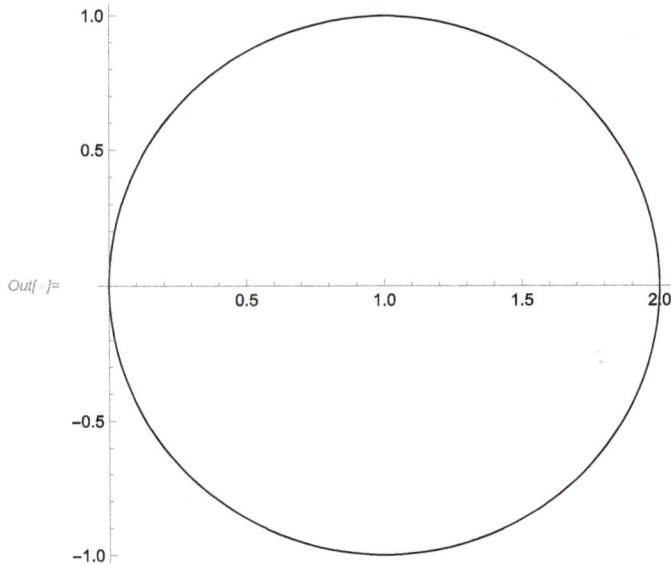

Figure 1.9

By means of the function Show, we can combine everything together:

In[·]:= Show[Plot[x^2, {x, 0, 1}], ContourPlot[
        x^2 + y^2 == 1, {x, -1, 1}, {y, -1, 1},
        PlotRange -> Automatic], Graphics[Circle[{1, 0}, 1],
        Axes -> Automatic]]

Out[·]=

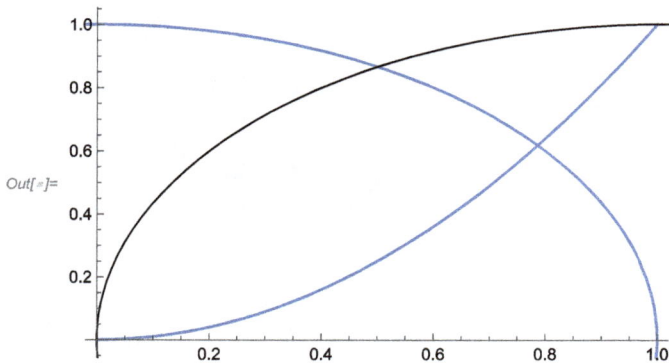

Figure 1.10

We can further use Dashing or colors (which are examples of so-called graphics Directives discussed below) to make our plots more attractive:

In[·]:= Graphics[{Dashing[{0.15, 0.20}], Red,
        Circle[{0, 0}, 1], AspectRatio -> Automatic}]

Out[·]=

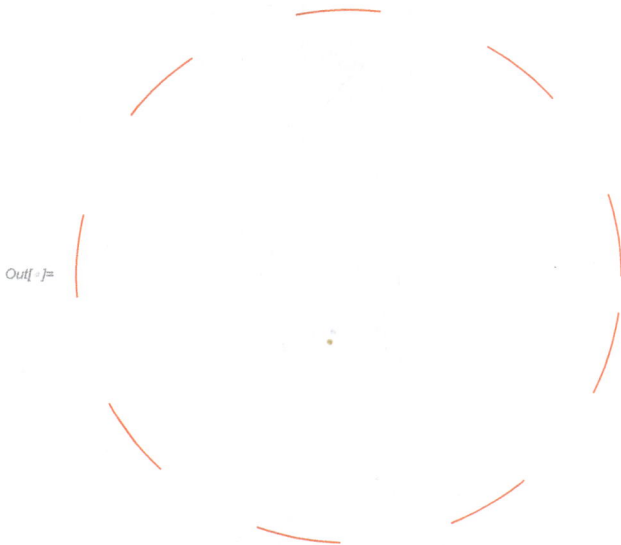

Figure 1.11

Let us give a few more examples of graphics primitives and the way we can color them:

```
In[·]:= Graphics3D[{RGBColor[0.3, 0.3, 0.7],
        Arrow[{{0, 0, 0}, {1, 1, 1}}]}]
```

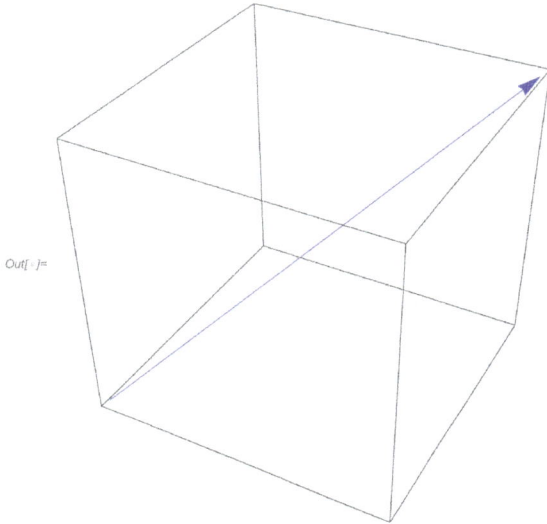

Out[·]=

Figure 1.12

```
In[·]:= Graphics3D[{Hue[0.8], Arrow[{{0, 0, 0},
        {1, 1, 1}}]}]
```

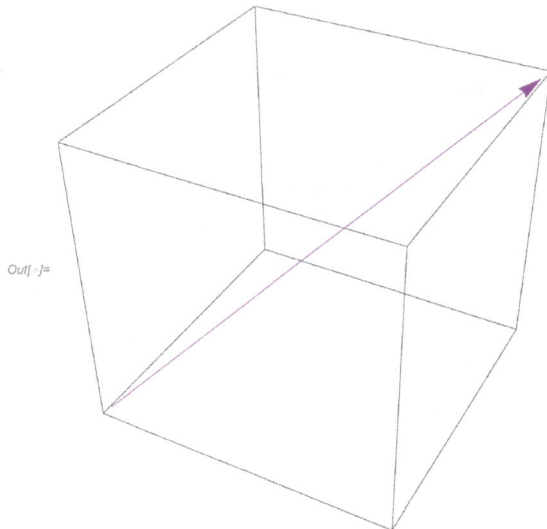

Out[·]=

Figure 1.13

We can make the arrow look three-dimensional by using Tube:

In[·]:= Graphics3D[{Red, Arrow[Tube[{{0, 0, 0},
　　{1, 1, 1}}, 0.01]], Green, Arrow[
　　Tube[{{0, 0, 0}, {-1, 1, 1}}, 0.01]]},
　　AspectRatio -> 1]

Out[·]=

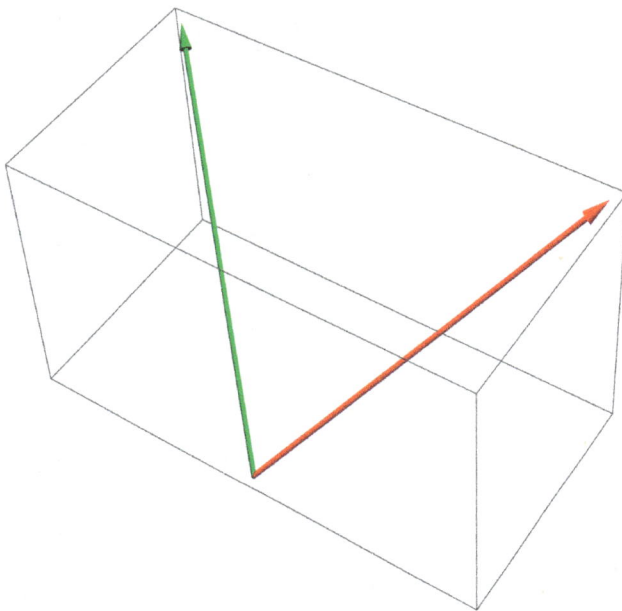

Figure 1.14

In general, 2D and 3D **Mathematica**® graphics are made of "primitives." Some primitives work in two and three dimensions, but some work only in two dimensions and some only in three dimensions. Point, Line, Polygon, Arrow, Text, and Simplex are examples of primitives that work in both two and three dimensions. Circle, Disk, Rectangle, and Raster are examples that work only in two dimensions. Cuboid, Sphere, Cone, Cylinder, and Tube are examples that work only in three dimensions. A primitive is always followed by a list of coordinates or a list of lists of coordinates. Some primitives require only one argument, but some take two. Most work with zero arguments. For instance, to draw a point (or points), we use

In[·]:= Graphics[Point[{2, 3}], Axes -> True,
　　ImageSize -> Small]

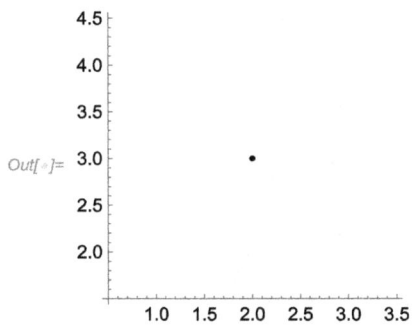

Out[ ]=

Figure 1.15

In[·]:= Graphics3D[Point[{2, 3, 4}], ImageSize -> Small]

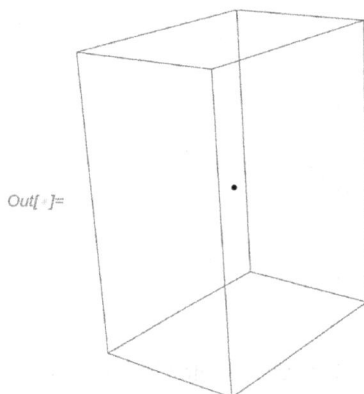

Out[ ]=

Figure 1.16

In[·]:= Graphics[Point[{{1, 1}, {-1, -1}}],
        Axes -> True, ImageSize -> Small]

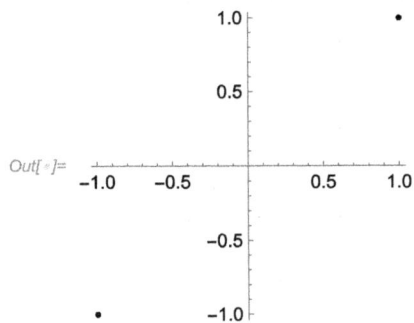

Out[ ]=

Figure 1.17

In[·]:= Graphics[{Circle[{0, 1}, 1], Point[
{1/2, 1/2}]}, Axes -> True]

Out[·]=

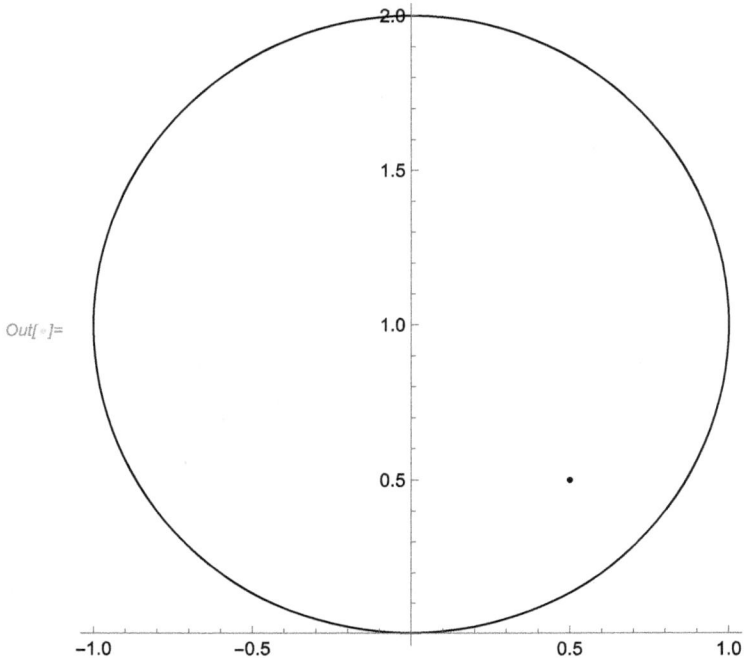

Figure 1.18

When graphics are built by means of graphic primitives, black color is used by default, and thickness and other properties are also decided by **Mathematica**®'s own algorithms. These can be changed by means of so-called Directives, which determine various properties of graphic objects. Here are some examples of directives which work with both Graphics and Graphics3D primitives: Red, RGBColor, Opacity, Hue, PointSize, Thickness, AbsoluteThickness, Thick, Thin, and Dashing. They work for functions which are found after them, not before. Here is an example of using directives with Point, Line and Disk:

```
In[·]:= Manipulate[Graphics[{Red, PointSize[0.02],
        Point[{1, 1}], RGBColor[r, g, b], Thickness[u],
        Line[{{-1, -1}, {0, 0}, {1, 1}}],
        PointSize[p], Point[{{1, 0}, {0, 1}}],
        Opacity[w], Hue[v], Disk[{0, 0}, 1]},
        Axes -> True], {{p, 0.1, "p"}, 0.01, 1,
        Appearance -> "Labeled"}, {{r, 1,
        "red"}, 0, 1, Appearance -> "Labeled"},
        {{g, 0, "green"}, 0, 1, Appearance ->
        "Labeled"}, {{b, 0, "blue"}, 0, 1,
        Appearance -> "Labeled"}, {{u, 0.001,
        "line thickness"}, 0.001, 0.1,
        Appearance -> "Labeled"}, {{w, 0.5,
        "opacity of disk"}, 0, 1,
        Appearance -> "Labeled"}, {{v, 0.5,
        "hue"}, 0, 1, Appearance -> "Labeled"}]
```

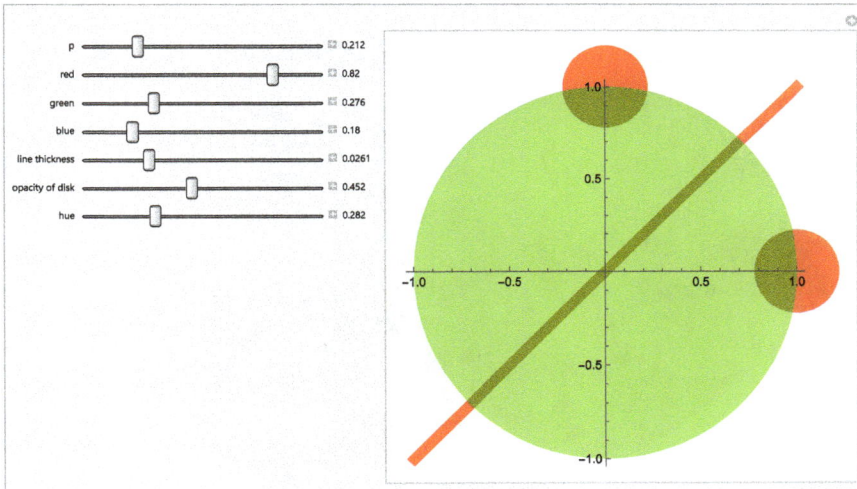

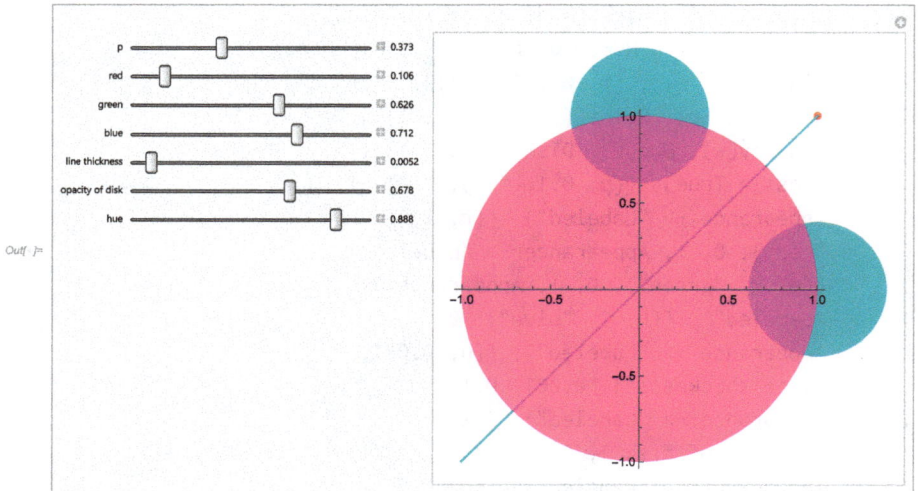

Figure 1.19

Three-dimensional plots are automatically colored and a default lighting is used.

```
In[·]:= parab = Plot3D[x^2 + y^2, {x, -1, 1},
        {y, -1, 1}]
```

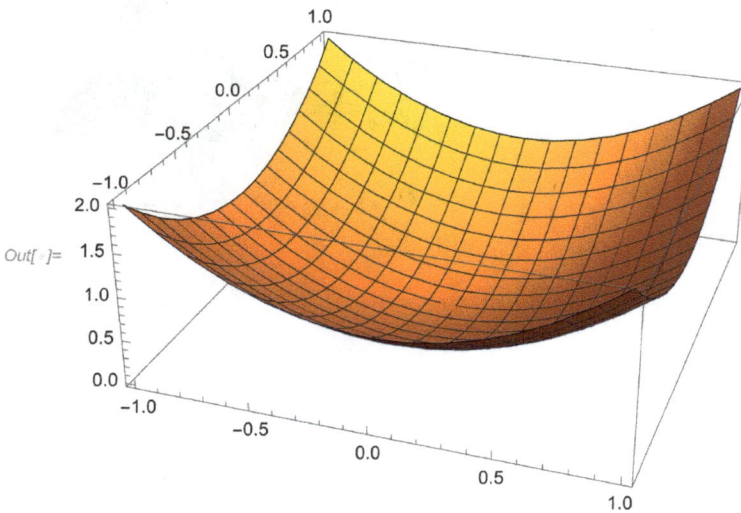

Figure 1.20

```
In[·]:= Plot3D[x^2 + y^2, {x, -1, 1}, {y, -1, 1},
        Lighting -> {{"Directional", Pink, {{5, 5, 4},
        {5, 5, 0}}}}]
```

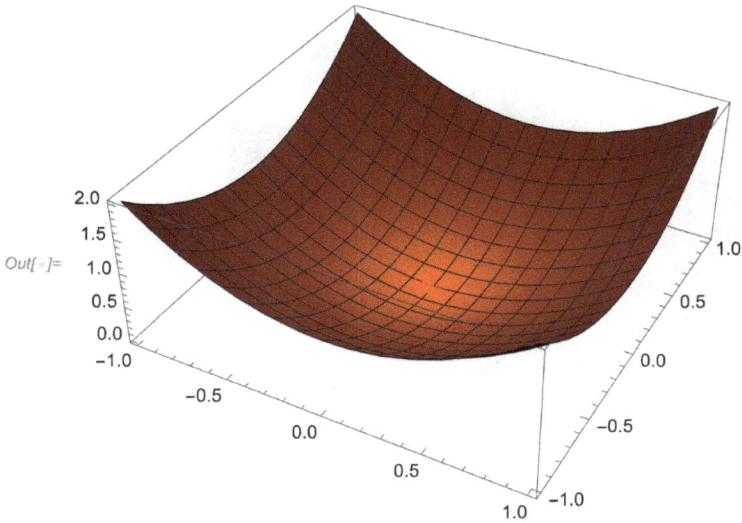

Figure 1.21

```
In[·]:= Plot3D[x^2 + y^2, {x, -1, 1}, {y, -1, 1},
        Mesh -> False, ColorFunction -> (Red &),
        Lighting -> "ThreePoint"]
```

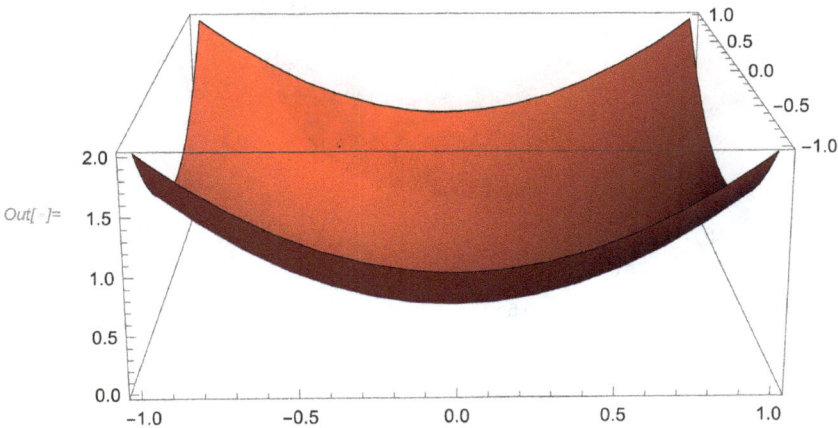

Figure 1.22

Note that we can change some features of a graphic object by using Show but not others:

```
In[·]:= Show[parab, Axes -> False, Boxed -> False]
```

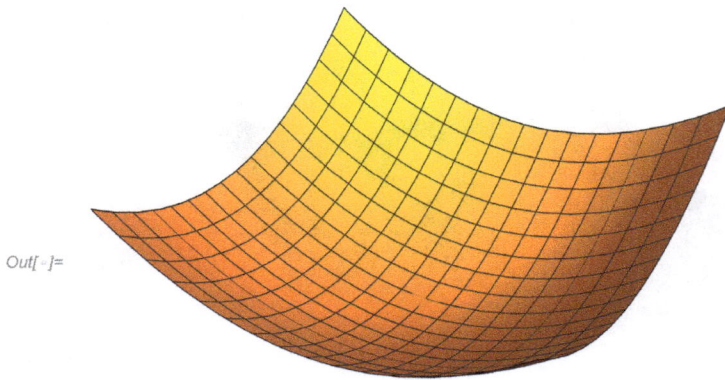

Out[ ]=

Figure 1.23

However,

In[·]:= Show[parab, Mesh -> None]

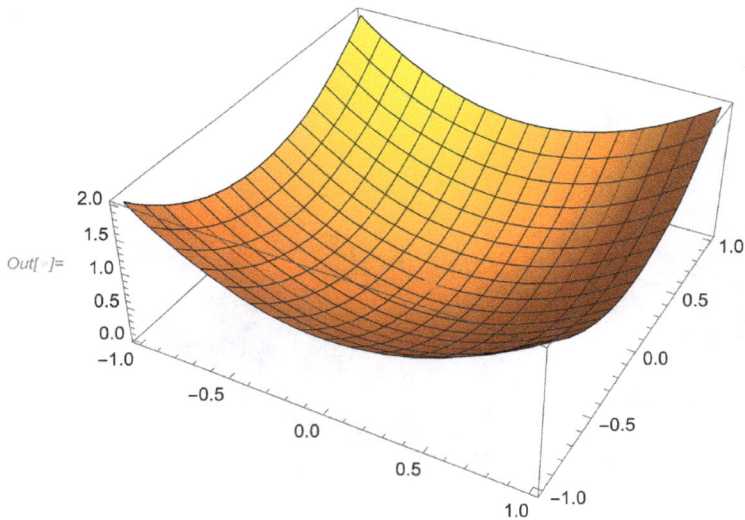

Figure 1.24

does not work. The reason is that

In[·]:= MemberQ[Options[Plot3D], Mesh,
        {1, Infinity}]

Out[·]:= True

In[·]:= MemberQ[Options[Graphics3D], Mesh,
{1, Infinity}]

Out[·]:= False

Thus,

In[·]:= Plot3D[x^2 + y^2, {x, -1, 1}, {y, -1, 1},
Mesh -> None]

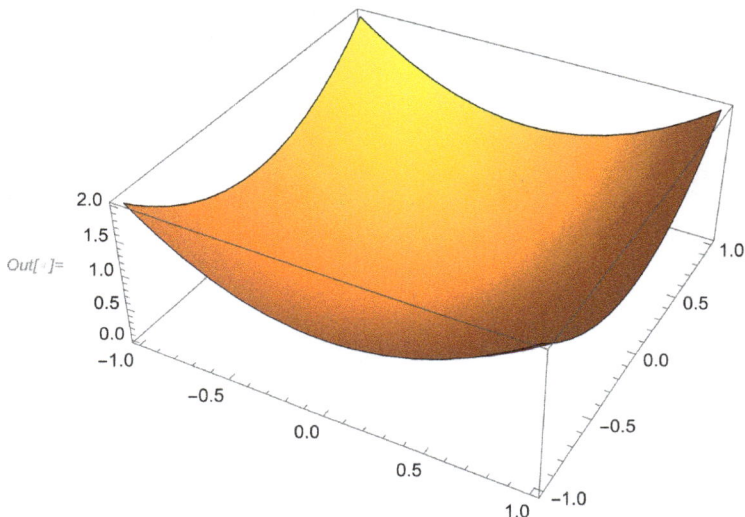

Figure 1.25

Using the function Manipulate, we can produce more complicated plots of functions
with parameters and regulate options like Mesh, Axes, Boxed, etc.

In[·]:= Manipulate[ParametricPlot3D[{Cos[v] +
(Sin[n*u] + Sin[20*v])/10, u, Sin[v]
+ (Cos[m*u] + Sin[20*v])/10}, {u, -Pi, Pi},
{v, -Pi, Pi}, PlotPoints -> 50, Mesh ->
mesh, Axes -> axes, Boxed -> boxed],
{n, 1, 5}, {m, 1, 5}, {mesh, {None,
Automatic, All}}, {axes, {False,
True}}, {boxed, {False, True}}]

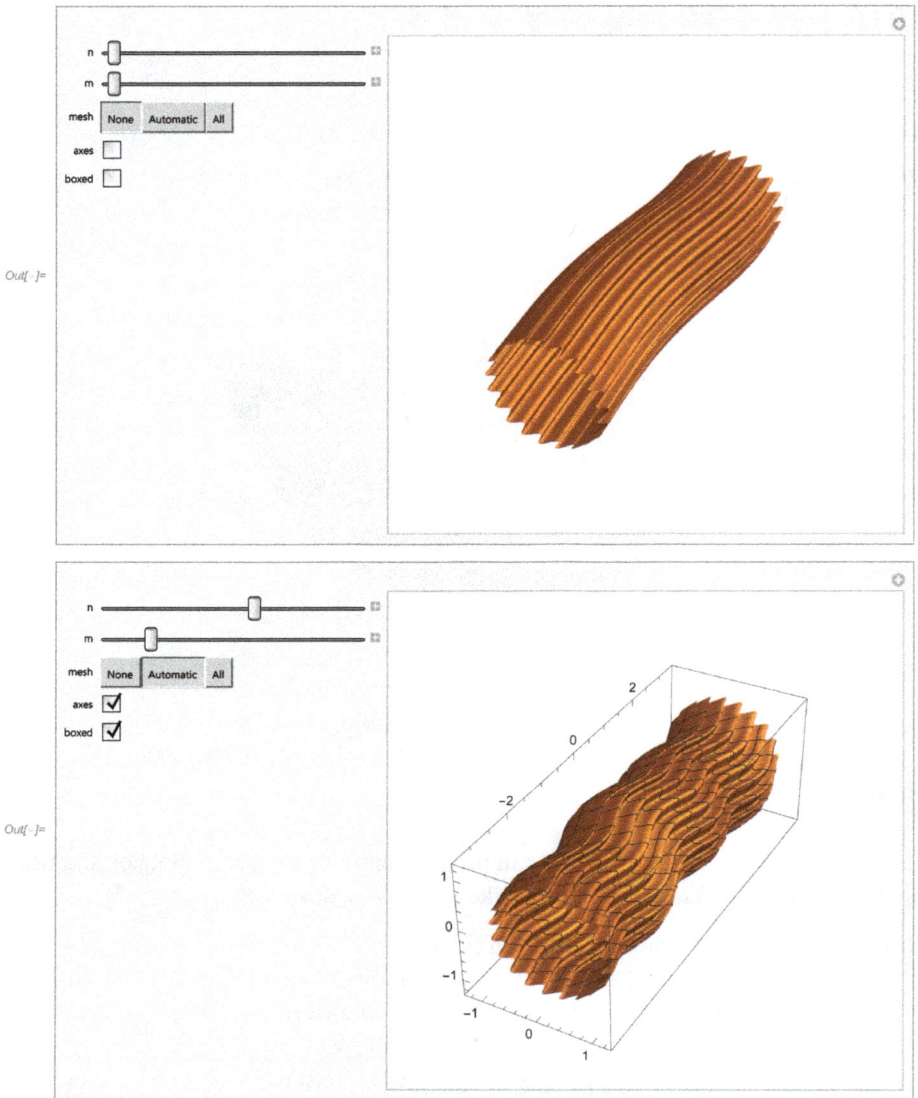

Figure 1.26

Plotting functions such as Plot, ParametricPlot, Plot3D, ParametricPlot3D, Region-Plot, RegionPlot3D, etc., also automatically constructs suitable Graphics or Graphics3D objects.

In[·]:= `Plot[{x^2, x^3}, {x, -1, 1}]`

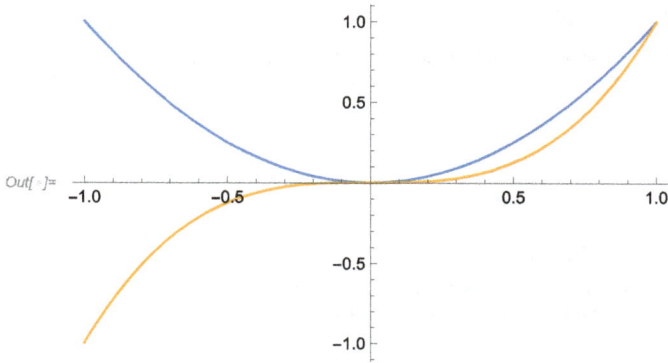

Figure 1.27

The graphics object is actually built up of primitives and directives. We can see it by converting it into `InputForm`, either using the menu or the Wolfram Language. Below we use `Short` to reduce to 5 lines the length of the code Mathematica® displays:

In[·]:= `Short[InputForm[%], 5]`

```
Graphics[{{{{}, {}, <<1>>, Annotation[{Directive[Opacity[1.],
   RGBColor[0.880722, 0.611041, 0.142051], AbsoluteThickness[1.6]], Line[{{-0.9999999591836735,
   -0.9999998775510254}, {-0.9993865641588807, -0.9981608211563976}, {-0.998773169134088,
   -0.9963240208976644}, {-0.9975463790845025, -0.9926571832488733}, <<440>>, {0.9993135319588606,
   0.9979420092682064}, {0.9999999591836735, 0.9999998775510254}}]}]}}, {}}, {<<25>>}]
```

Figure 1.28

Note that the `Plot` is made of lists of points joint by lines, which can be seen by converting to `InputForm` as before.

In[·]:= `parabola = Plot[x^2, {x, -1, 1}]`

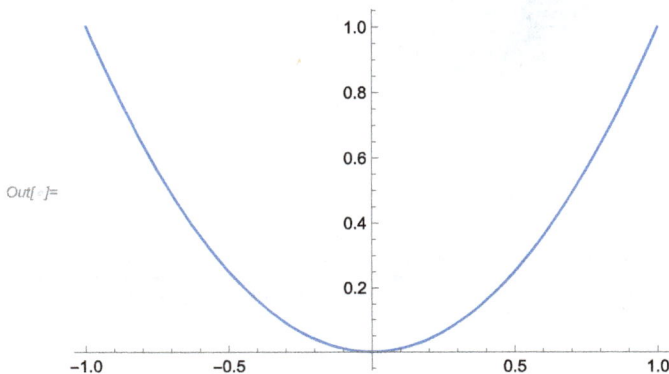

Figure 1.29

We can see a few (2 in this case) lines of the InputForm:

In[·]:= Short[InputForm[parabola], 2]

Out[·]//Short= Graphics[{{{{}, {}, Annotation[{Directive[Opacity[1.], RGBColor[0.368417,
0.506779, 0.709798], AbsoluteThickness[1.6]], Line[<<1>>]}, <<1>>]}]}, {}}, {<<25>>}]

Figure 1.30

For the paraboloid parab defined above,

In[·]:= Short[InputForm[parab], 2]

Out[·]//Short= Graphics3D[{GraphicsComplex[{{-0.9999998571428571, -0.9999998571428571,
1.9999994285714693}, <<1364>>}, {0.12499998214285706, <<2>>}}, <<2>>], {}, {<<15>>}]

Figure 1.31

To see the objects in the graphics, we use Normal

In[·]:= Short[InputForm[Normal[parab]], 2]

Out[·]//Short= Graphics3D[{{{{EdgeForm[], Directive[<<3>>], GraphicsGroup[Annotation[{{Polygon[{<<3>>},
<<1>>], <<711>>}, <<2>>, {<<3>>}}, <<1>>]}]}, <<4>>}, {<<5>>}}, {}, {<<15>>}]

Figure 1.32

We can also see the InputForm as follows:

In[·]:= **InputForm[**

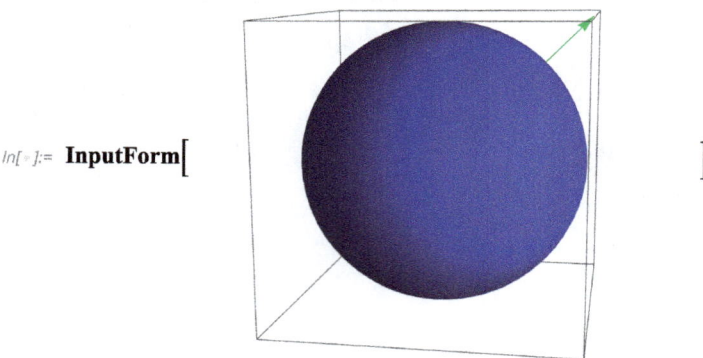

]

Figure 1.33

Out[·]//InputForm= Graphics3D[{{RGBColor[1, 0, 0], Arrow[{{0, 0, 0}, {1, 1, 1}}]}}, {RGBColor[0, 1, 0], Arrow[{{0, 0, 0}, {1, 1, 1}}]}},
{RGBColor[0, 0, 1], Sphere[{0, 0, 0}]}}, AspectRatio -> 1, ImageSize -> {275.4592738156954, 293.74533260039254},
ImageSizeRaw -> Automatic, ViewPoint -> {0.5824816563389832, -3.2754903959787267, 0.6179626088038986},
ViewVertical -> {0.08376968168439568, -0.4257195903462696, 0.900969184171079}}]

Figure 1.34

This means that we can operate on a graphic in the same way as on any **Mathematica**® expression, for example, we can perform replacements:

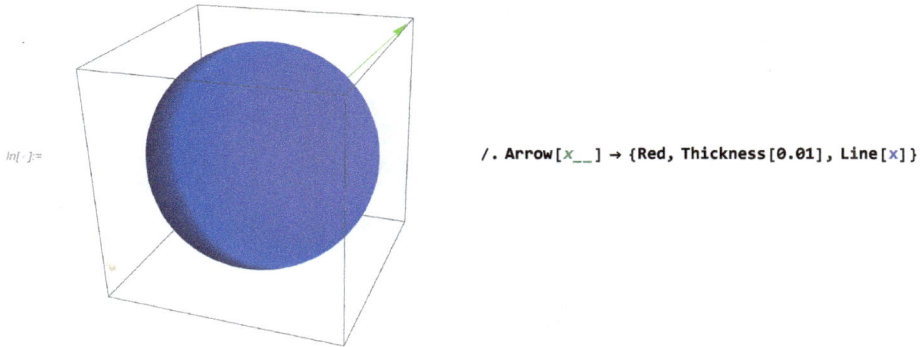

In[ ]:=   `/. Arrow[x__] → {Red, Thickness[0.01], Line[x]}`

Figure 1.35

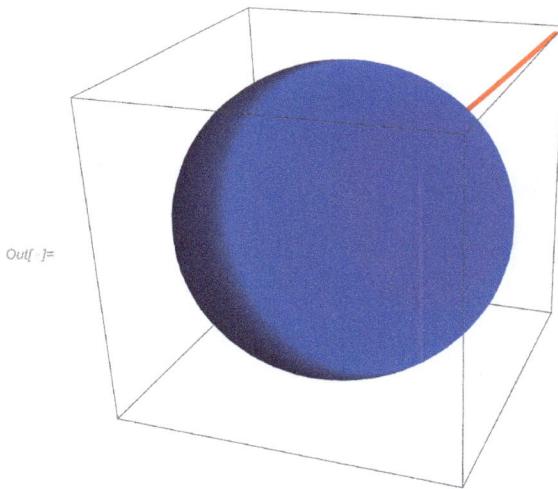

Out[ ]=

Figure 1.36

The green arrow in the first picture was replaced by the red line in the second one.

Similarly, we can replace in sphere created by `ParametricPlot3D` all polygons which form this sphere by lines.

```
In[·]:= ParametricPlot3D[{Cos[u] Cos[v],
        Cos[v] Sin[u], Sin[v]}, {u, 0, Pi},
        {v, -Pi, Pi}, ColorFunction ->
        Function[{x, y, z}, Blue]]
```

Out[ ]=

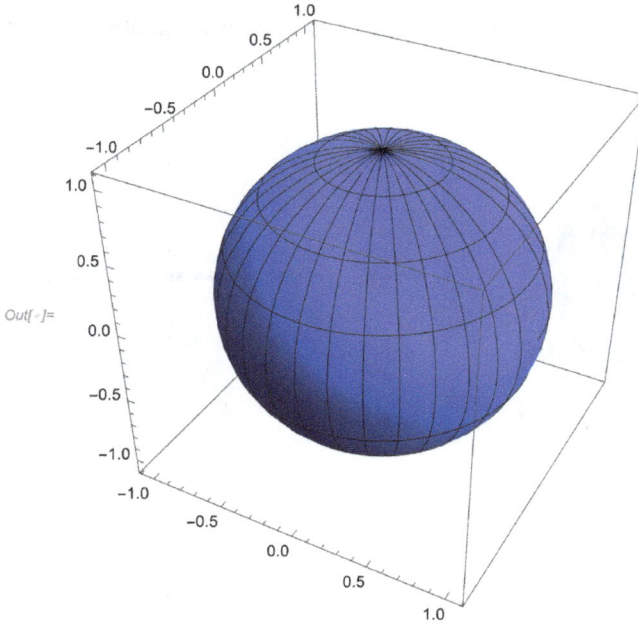

Figure 1.37

In[·]:= % /. Polygon -> Line

Out[ ]=

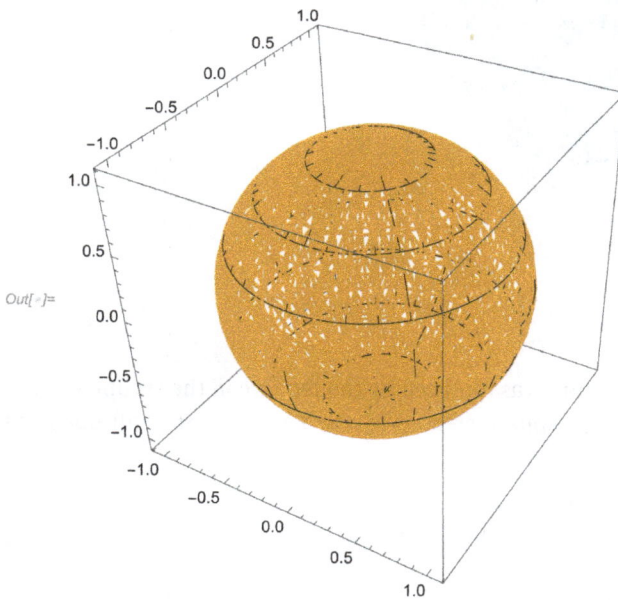

Figure 1.38

As mentioned before, graphics of the same type can be combined. There are several ways to do this. Here is one example:

In[·]:= Graphics[{Red, Opacity[0.5], Disk[{0, 1},
        1], Blue, Disk[{1, 0}, 1]}]

Out[·]=

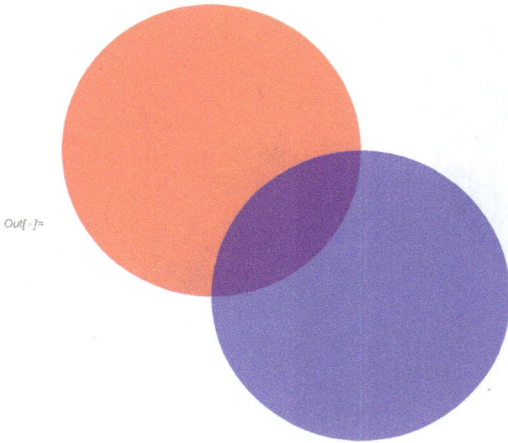

Figure 1.39

Observe that, due to the transparency of the blue circle, we can see the red one behind it. The same can be achieved by making two independent pictures and combining them using the command Show.

In[·]:= disk1 = Graphics[{Red, Opacity[0.5],
        Disk[{0, 1}, 1]}]

Out[·]=

Figure 1.40

In[·]:= `disk2 = Graphics[{Blue, Disk[{1, 0}, 1]}]`

Out[·]=

Figure 1.41

In[·]:= `Show[disk2, disk1]`

Out[·]=

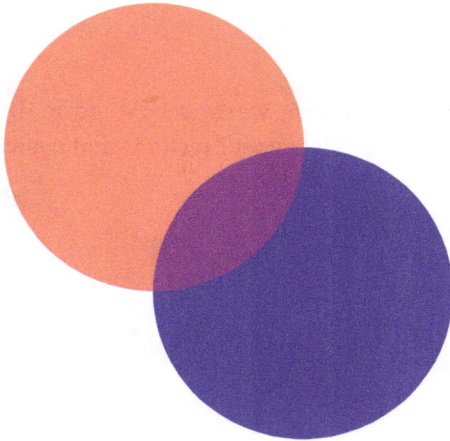

Figure 1.42

Note that the order of objects matters:

In[·]:= `Show[disk1, disk2]`

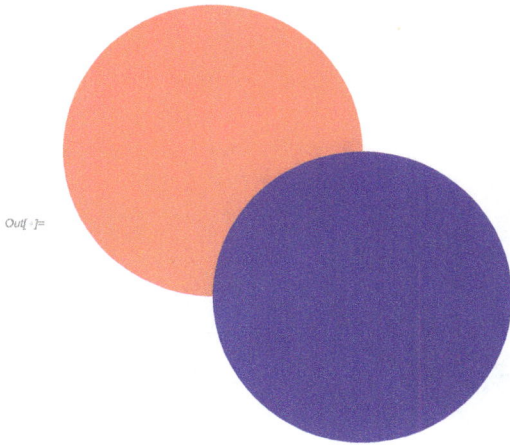

Out[ ]=

Figure 1.43

Show is particularly useful when combining a graphics object made by using primitives with that made by a plotting function. For example,

In[·]:= sphere = Graphics3D[{Opacity[0.2], Red,
        Sphere[{0, 0, 0}, 1]}]

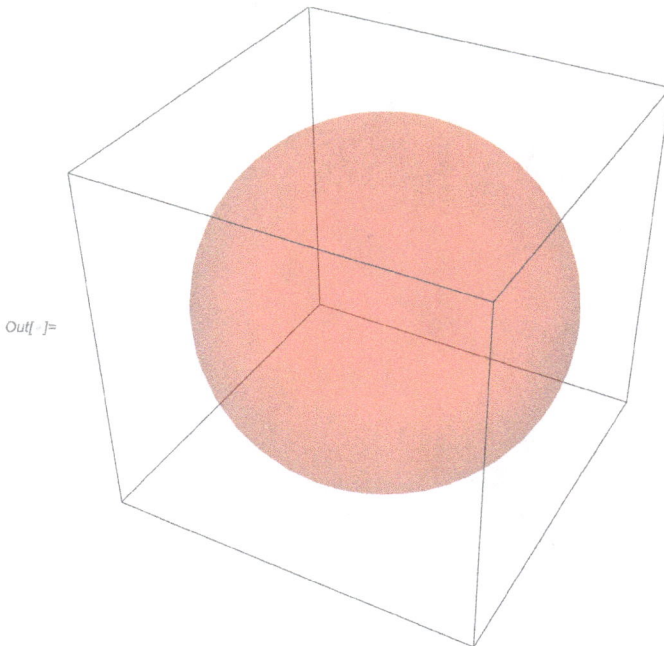

Out[ ]=

Figure 1.44

In[·]:= `parab1 = Plot3D[x^2 + y^2, {x, -1, 1}, {y, -1, 1},`
`    RegionFunction -> Function[{x, y, z},`
`    x^2 + y^2 + z^2 <= 1]]`

Out[·]=

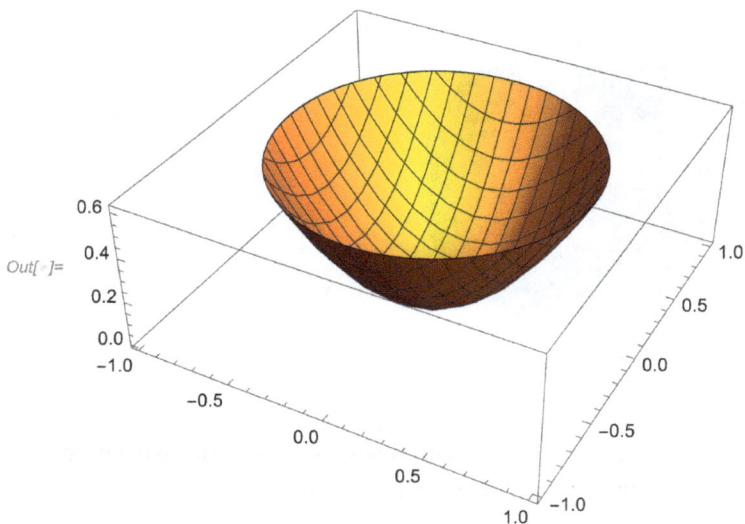

Figure 1.45

In[·]:= `Show[sphere, parab1, Boxed -> False]`

Out[·]=

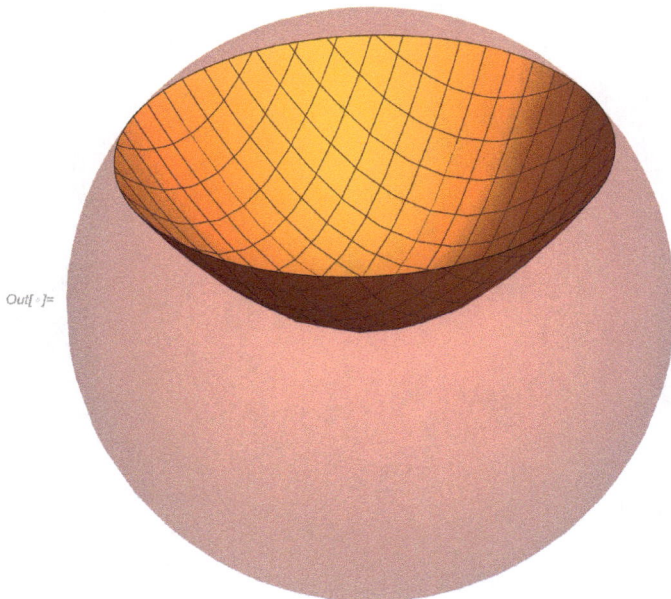

Figure 1.46

The picture above does not look three dimensional, so it is better to omit `Boxed->False` (same as setting `Boxed->True`):

In[·]:= `Show[sphere, parab1]`

Out[·]=

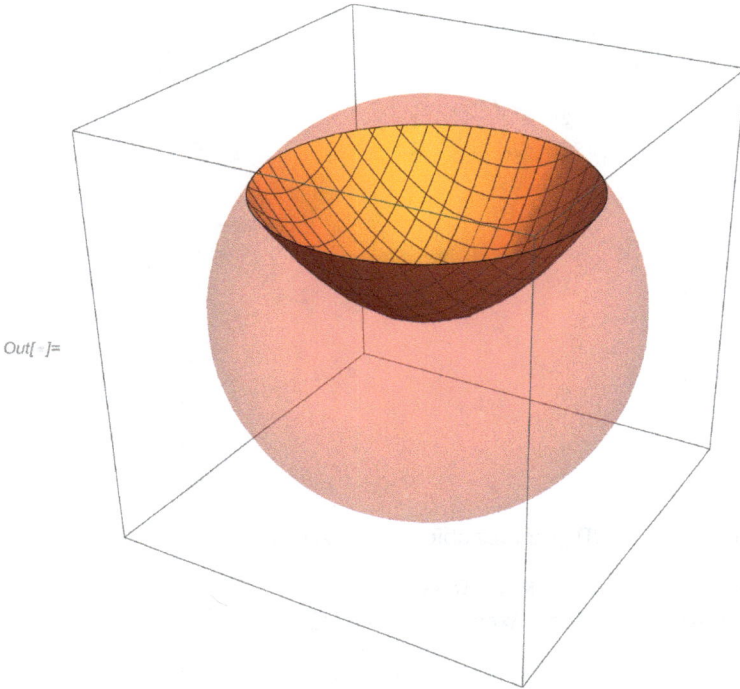

Figure 1.47

Finally, one more useful way to combine graphics is by using the function `Inset`. It can be used to combine 2D and 3D graphics.

In[·]:= `Graphics[{LightBlue, Disk[], Inset[`
        `Plot[Tan[x], {x, -3, 3}]]}]`

Out[ ]=

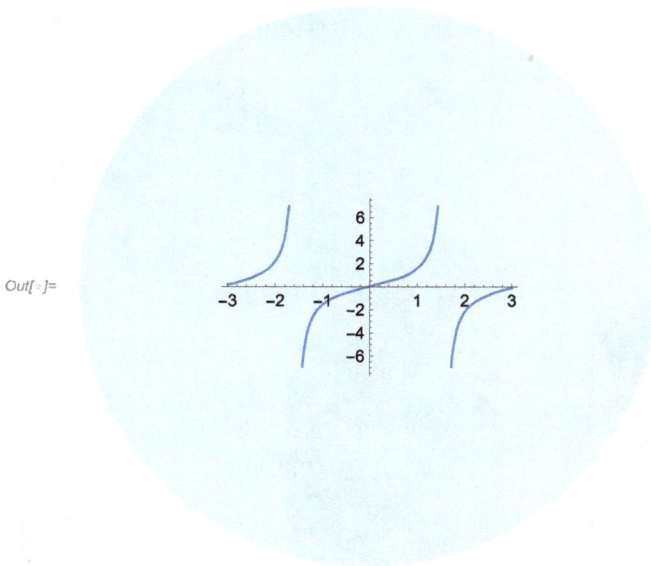

Figure 1.48

Inset can be used to insert a 3D graphics object into a 2D one:

In[ ]:= Graphics[{LightBlue, Disk[], Inset[Show[
       parab1, Boxed -> False, Axes -> False]]}]

Out[ ]=

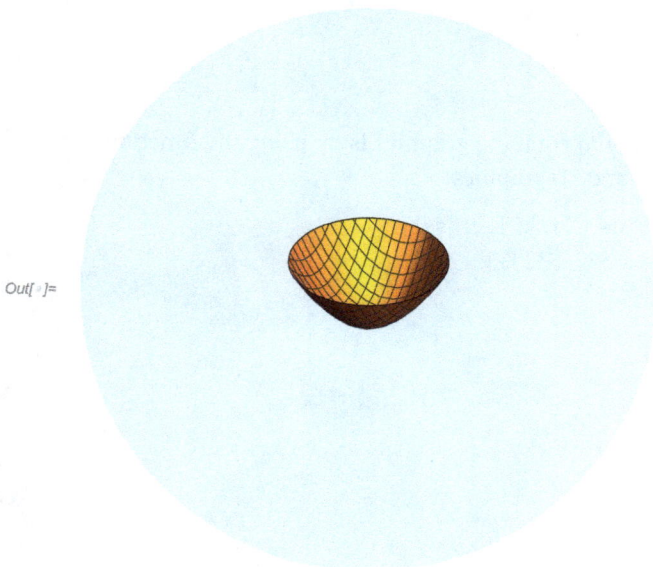

Figure 1.49

Conversely, we can inset a 2D object into a 3D object:

```
In[·]:= Graphics3D[{Cuboid[{-0.25, 0.5, -0.25},
        {0.25, 1, 0.25}], Inset[Graphics[{Disk[
        {0, 0}, .25]}, ImageSize -> 50], {0, 0, 0}]},
        PlotRange -> 2]
```

Out[ ]=

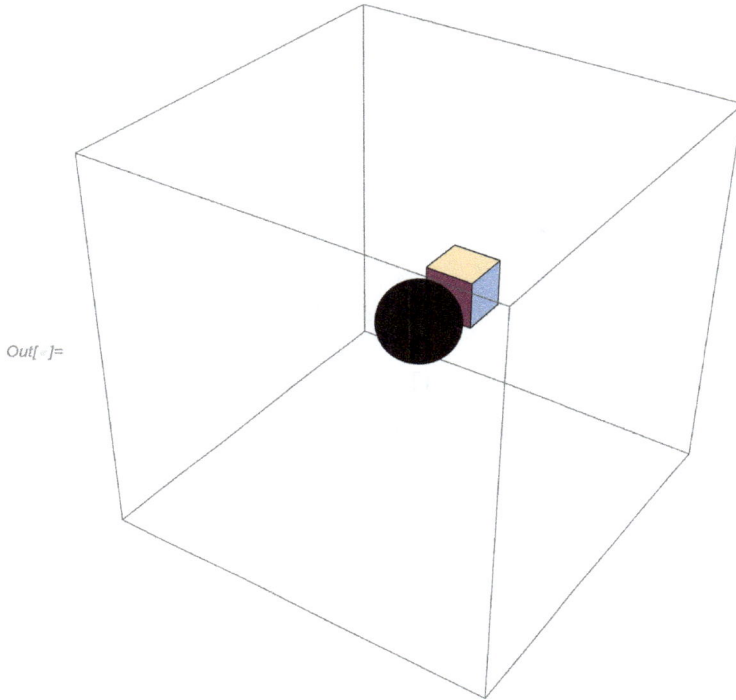

Figure 1.50

Although we shall not use this function often, it is still useful to know.

In Mathematica® 12.2 a new useful function Canvas appeared. It represents an empty canvas in the current notebook in which one can do some free-form drawing. This is useful if we want add schematically, for instance, some text or explanation to a picture, or draw some more arrows, lines, or a circle, etc. The advantage is that one does not need to specify coordinates where such objects should be placed.

```
In[·]:= Canvas[Plot[Sin[x], {x, -1, 1}]]
```

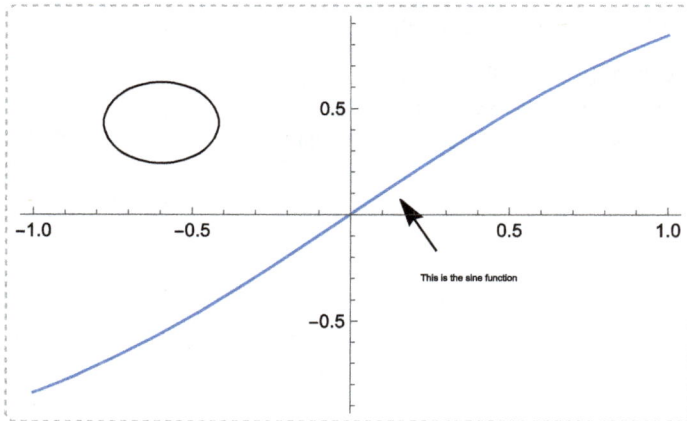

Figure 1.51

Numerical data

In[·]:= data = Table[Random[Real, {-1, 1}], {i, 1, 20}];

can be visualized with functions like ListPlot or others

In[·]:= ListPlot[data]

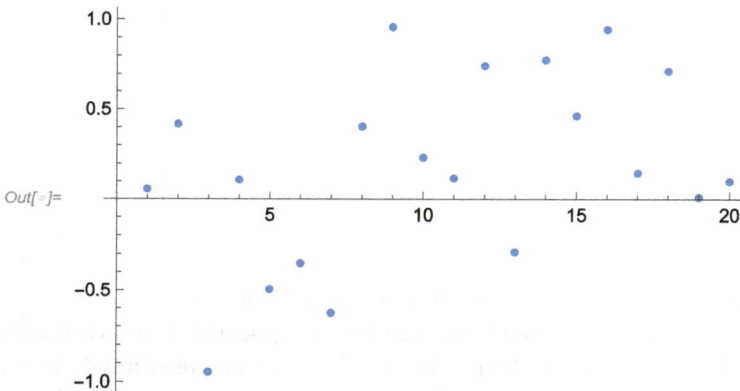

Figure 1.52

In[·]:= ListLinePlot[data]

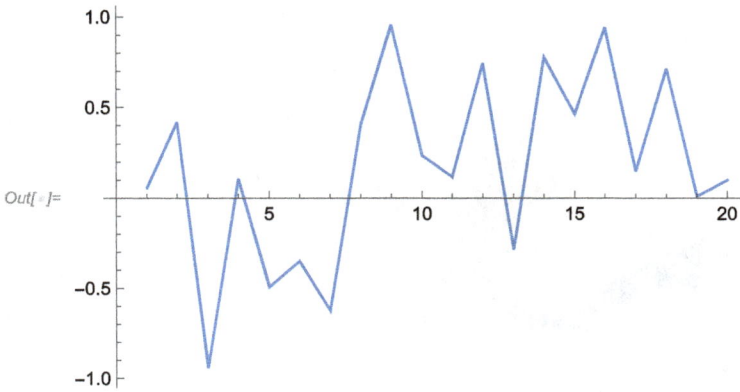

Figure 1.53

```
In[·]:= data3D = Table[{Random[Real, {-1, 1}],
        Random[Real, {1, 10}], Random[
        Real, {0, 10}]]}, {i, 1, 10}];
```

```
In[·]:= ListPointPlot3D[data3D]
```

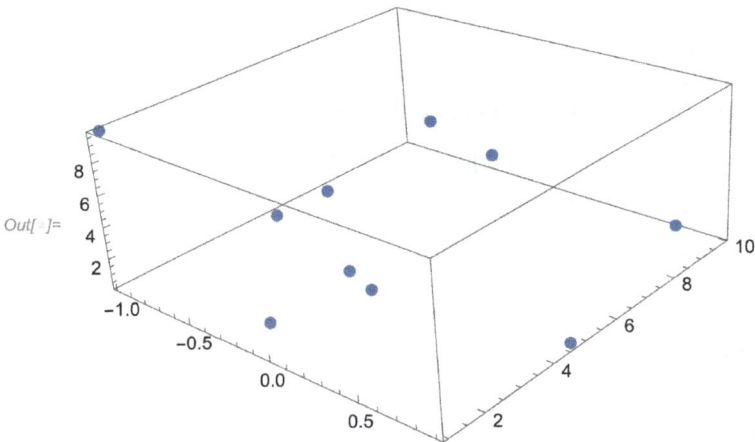

Figure 1.54

```
In[·]:= ListPlot3D[data3D]
```

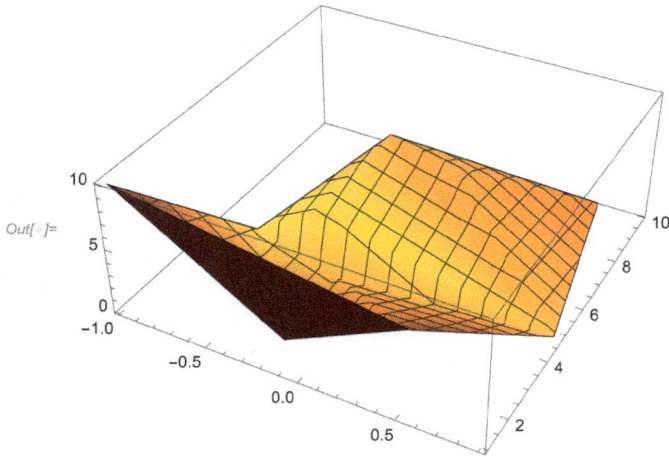

Out[ ]=

Figure 1.55

By typing in (without ;)

In[·]:= PolyhedronData[];

In[·]:= Length[%]
Out[·]:= 214

we see that **Mathematica**® can draw a lot of built-in polyhedra, for instance,

In[·]:= PolyhedronData["Icosahedron"]

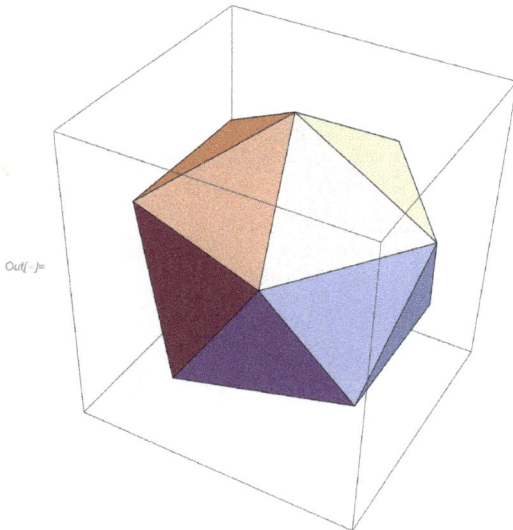

Out[ ]=

Figure 1.56

We should remark that the function Evaluate may prove useful when using the built-in graphics function. Compare the following outputs:

In[·]:= Plot[RandomReal[], {x, 0, 1}]

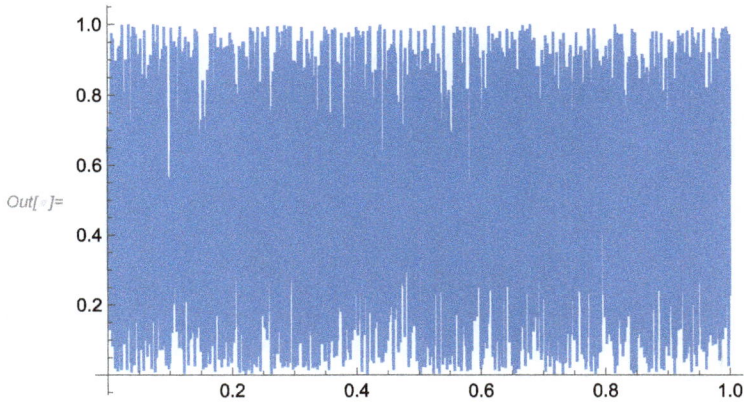

Figure 1.57

In[·]:= Plot[Evaluate[RandomReal[]], {x, 0, 1}]

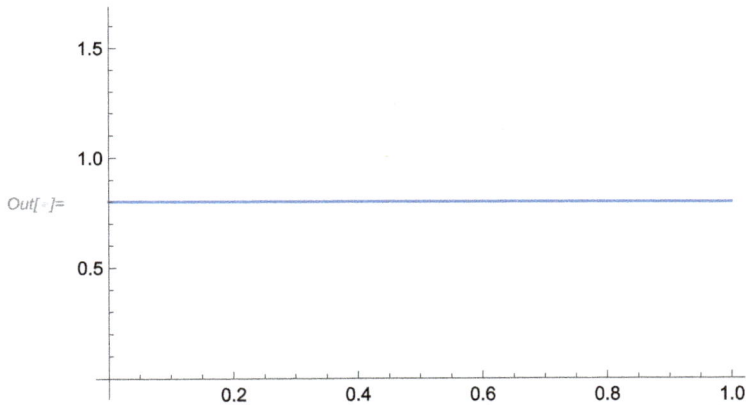

Figure 1.58

In[·]:= Plot[Table[BesselJ[n, x], {n, 4}],
        {x, 0, 10}, Filling -> Axis]

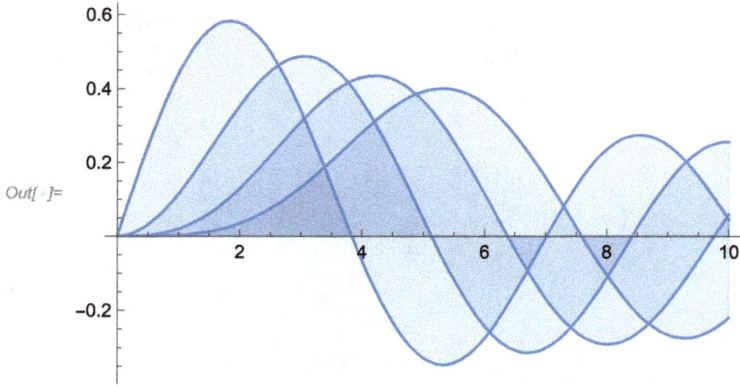

Figure 1.59

In[·]:= Plot[Evaluate[Table[BesselJ[n, x], {n, 4}]],
    {x, 0, 10}, Filling -> Axis]

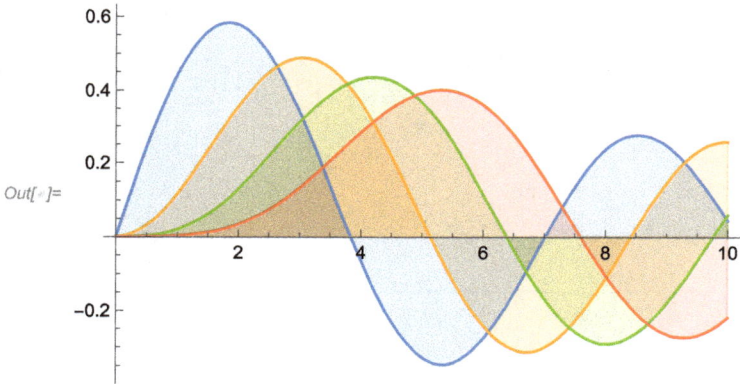

Figure 1.60

In the following example, we see that **Mathematica**® distinguishes two functions (a list of two elements in the function listf) if we use Evaluate when plotting it:

In[·]:= listf[x_] := {Sin[x], Cos[x]}

In[·]:= Plot[Evaluate[listf[x]], {x, -1, 1}]

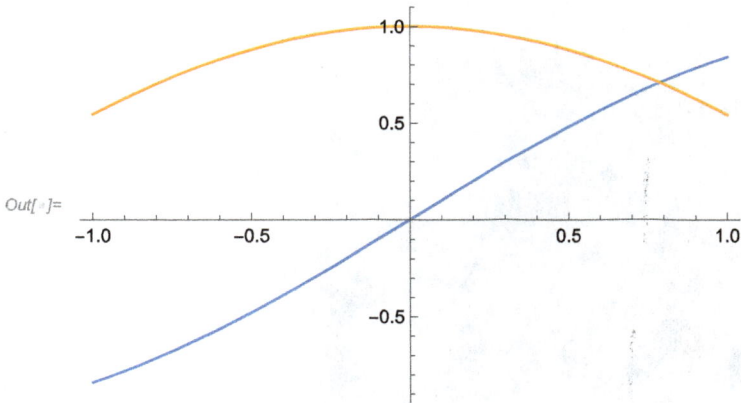

Figure 1.61

however, it colors everything in blue (so it does not distinguish two graphs) if Evaluate
is not used:

In[ ]:= Plot[listf[x], {x, -1, 1}]

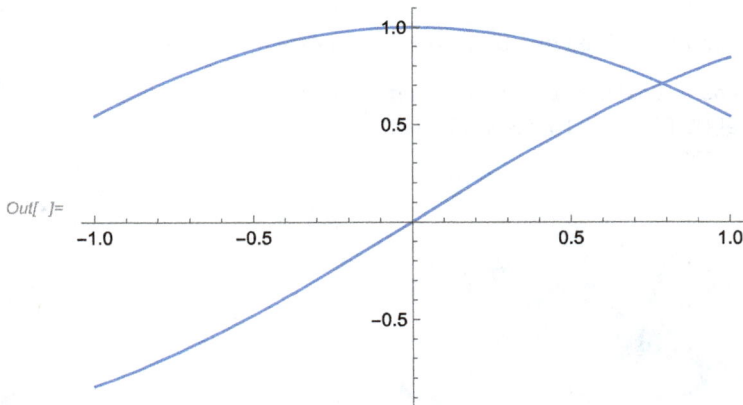

Figure 1.62

One more example is

In[ ]:= grcos[x, y] := Cos[x y]

In[ ]:= Plot3D[Evaluate[grcos[x, y]], {x, 0, 10}, {y, 0, 10}]

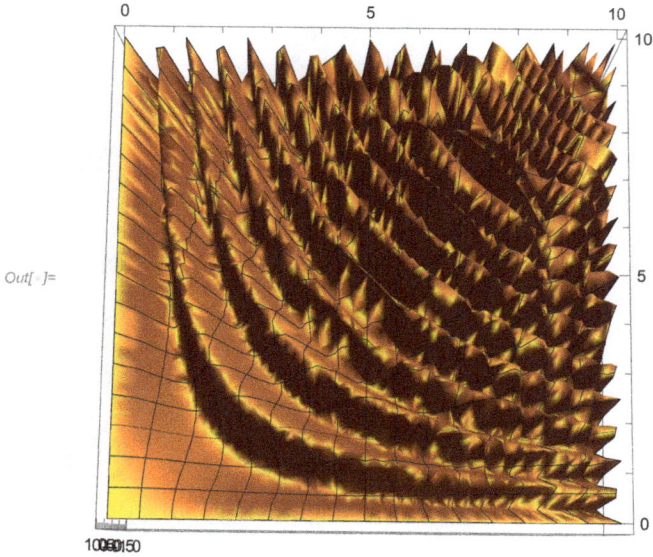

Out[ ]=

Figure 1.63

We can visualize complex-valued functions to a certain extent, for instance,

```
In[ ]:= Block[{fcos = Cos[x + I y]}, ParametricPlot[
         Evaluate[{Re[fcos], Im[fcos]}], {x, -Pi, Pi},
         {y, -2, 2}]]
```

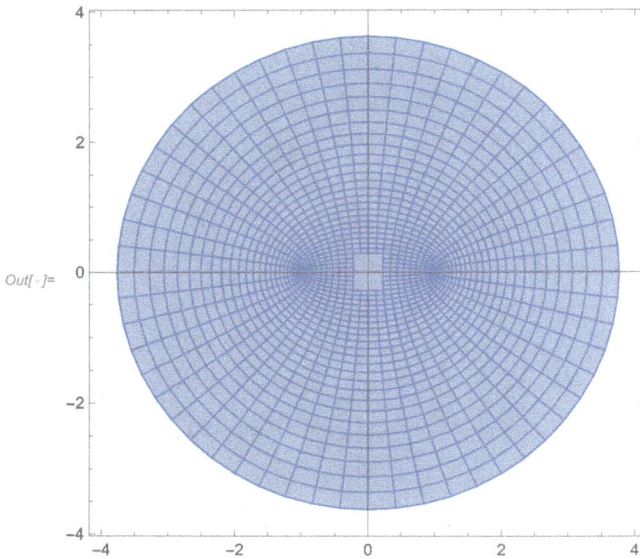

Out[ ]=

Figure 1.64

In[·]:= `Manipulate[Block[{fF = F[x + I y]},`
`        ParametricPlot[Evaluate[{Re[fF], Im[fF]}],`
`        {x, -Pi, Pi}, {y, -2, 2}]], {F, {Sin, Tan, Cot}}]`

Out[ ]=

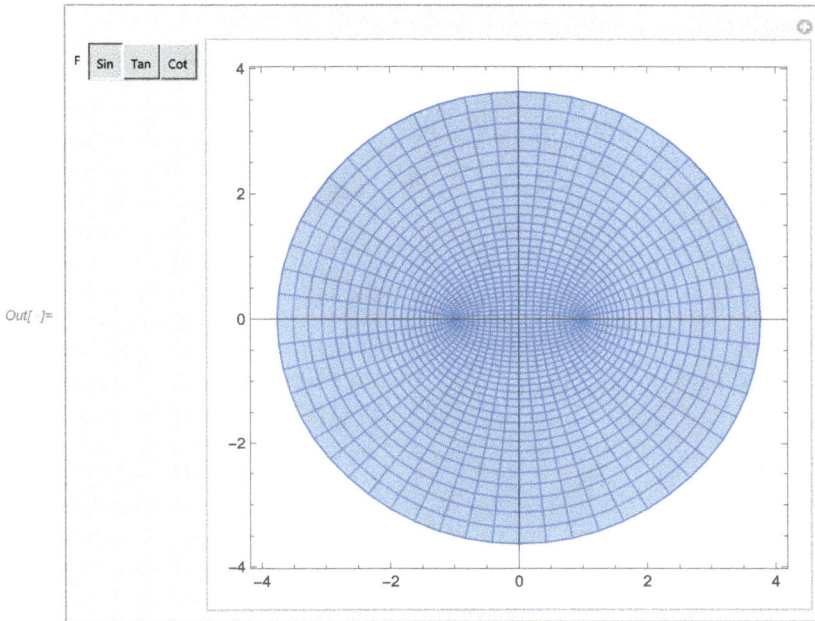

Figure 1.65

The function `ReImPlot` evaluates the function f at different values of the independent variable x to create smooth curves of the form $(x, \mathrm{Re}\, f(x))$ and $(x, \mathrm{Im}\, f(x))$.

In[·]:= `ReImPlot[Sqrt[(x^2 - 1)*(x^2 - 4)], {x, -3, 3}]`

Out[ ]=

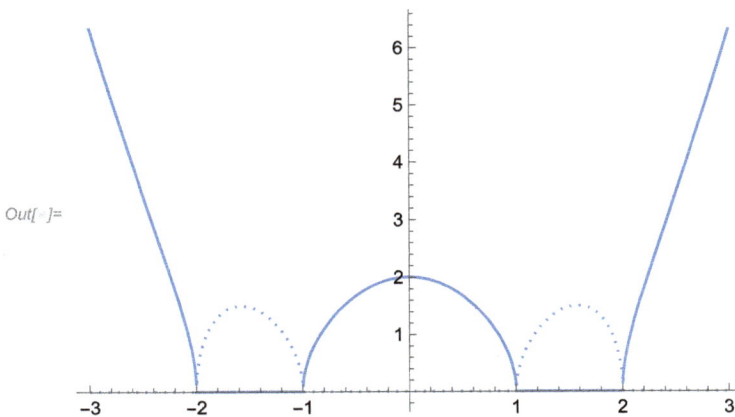

Figure 1.66

Compare with

In[·]:= `Grid[{{Plot[Re[Sqrt[(x^2 - 1)*(x^2 - 4)]],`
`{x, -3, 3}], Plot[Im[Sqrt[(x^2 - 1)*`
`(x^2 - 4)]], {x, -3, 3}]}}]`

Out[·]=

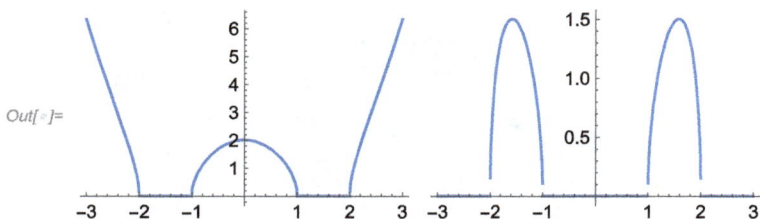

Figure 1.67

In[·]:= `ReImPlot[{ArcSin[x], ArcCos[x]}, {x, -4, 4}]`

Out[·]=

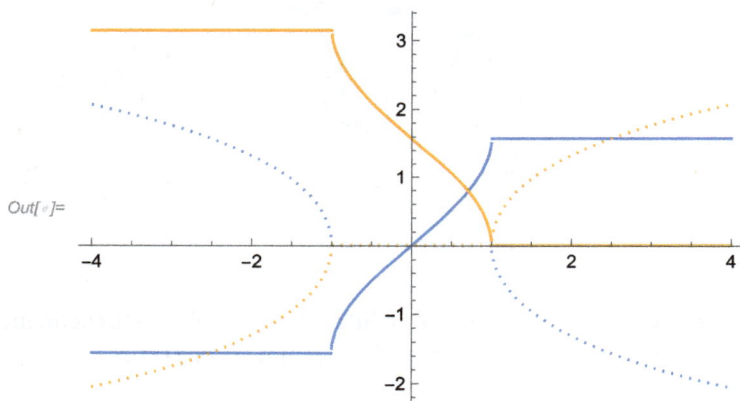

Figure 1.68

The function `ComplexPlot` generates a plot of $\text{Arg} f$ over the complex rectangle with corners z_{min} and z_{max}:

In[·]:= `ComplexPlot[(z^2 + 1)/(z^2 - 1), {z,`
`-2 - 2*I, 2 + 2*I}]`

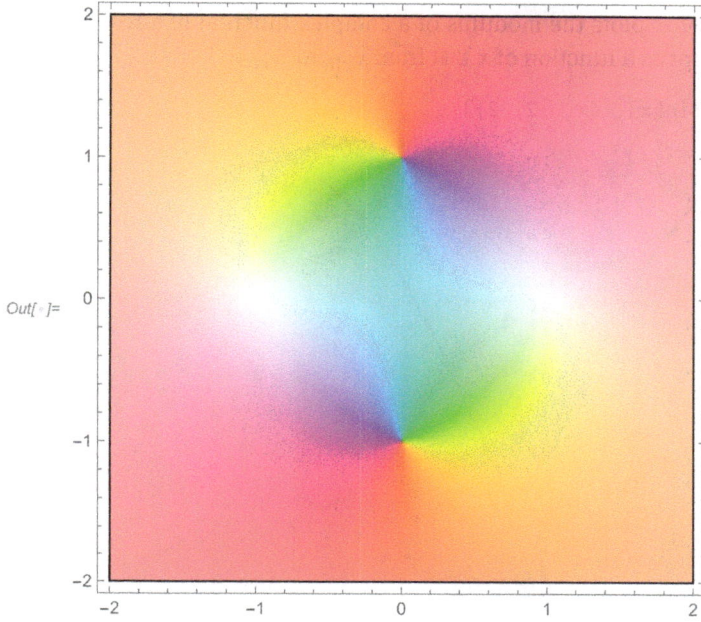

Figure 1.69

In the next picture we can see the poles and zeros of the function

```
In[·]:= ComplexPlot3D[(z^2 + 1)/(z^2 - 1),
         {z, -2 - 2*I, 2 + 2*I}]
```

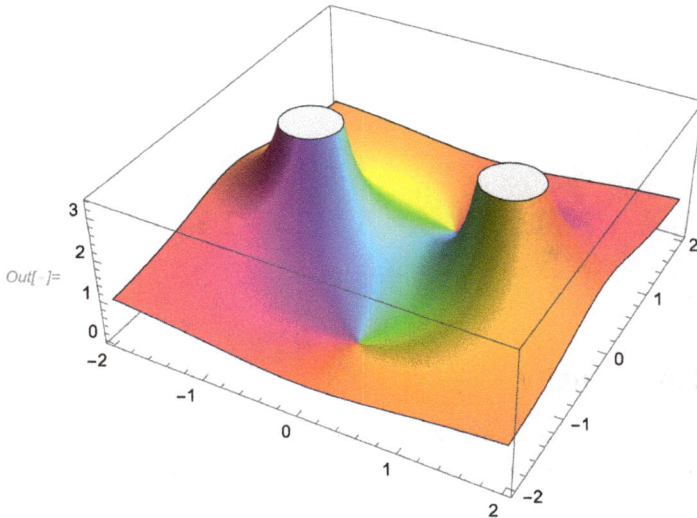

Figure 1.70

The function `AbsArgPlot` plots the modulus of a complex function of a real variable colored by its argument as a function of $x \in \mathbb{R}$ from x_{min} to x_{max}:

In[·]:= `AbsArgPlot[Sin[x], {x, -2, 2}]`

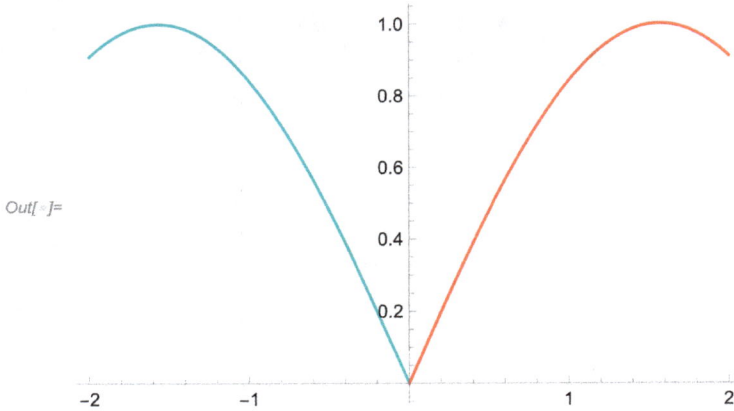

Figure 1.71

In[·]:= `AbsArgPlot[Sin[I x], {x, -2, 2}]`

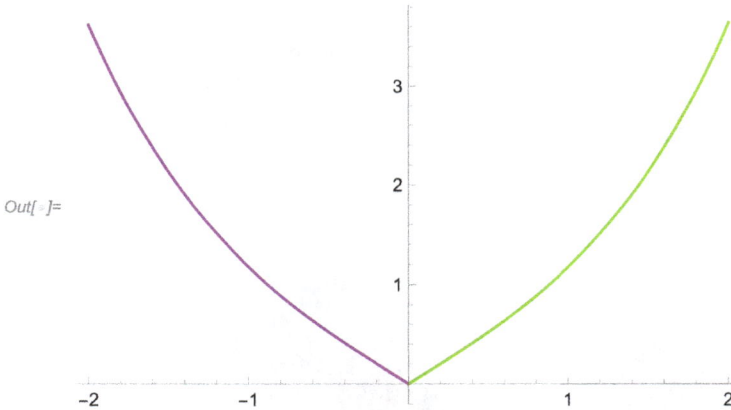

Figure 1.72

1.2 Example 1: Roots of unity

Let us draw the roots of unity in the complex plane (that is, solutions of $z^n = 1$) and show that as n tends to infinity they lie on the circle more and more densely.

In the following we shall use the function ReIm which gives a list of real and imaginary parts of the complex number z. This is equivalent to $\{Re[\#], Im[\#]\}\&$. The function ReIm is listable, as can be seen from its attributes, so it can be applied to lists.

In[·]:= Attributes[ReIm]

Out[·]:= {Listable, Protected}

Note that the function SetAttributes can be used to assign new attributes to other functions.

Let us define the function which draws roots of $z^n = 1$. The roots are sorted by their arguments and are joint by a line. We can also use a parameter r to scale the roots to lie on a circle of a bigger radius (by default r is taken to be equal to 1).

```
In[·]:= poly[n_, r_ : 1] := Graphics[{Red, Line[
        r*ReIm[Append[l = Sort[z /.
        Solve[z^n == 1, z], Arg[#1] <=
        Arg[#2] & ], First[l]]]]}]
```

```
In[·]:= Show[poly[11], poly[11, 2]]
```

Out[ ]=

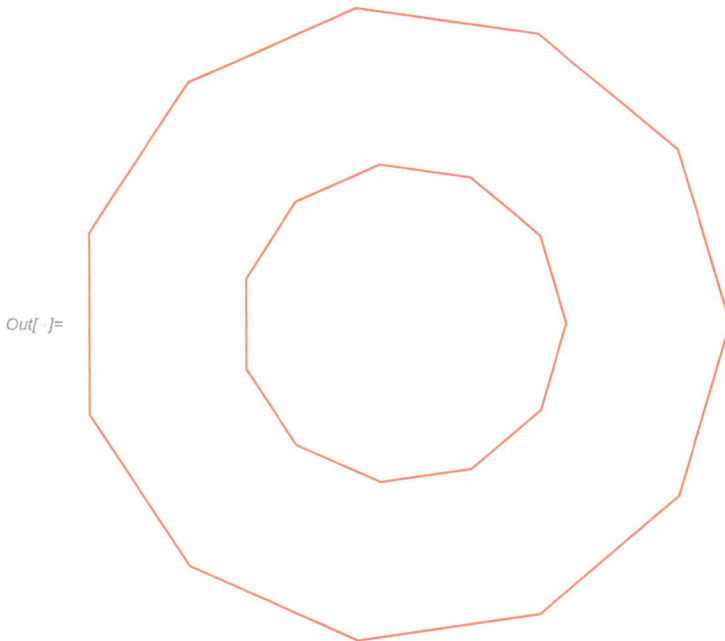

Figure 1.73

We can check the InputForm of poly[3]:

In[·]:= poly[3] // InputForm

Out[·]//InputForm=
```
Graphics[{RGBColor[1, 0, 0], Line[{{-1/2, -1/2*Sqrt[3]}, {1, 0}, {-1/2, Sqrt[3]/2}, {-1/2, -1/2*Sqrt[3]}}]}]
```

Figure 1.74

and can change the color by substitution

In[·]:= Show[poly[11] /. RGBColor[1, 0, 0] ->
        Blue, poly[11, 2]]

Out[·]=

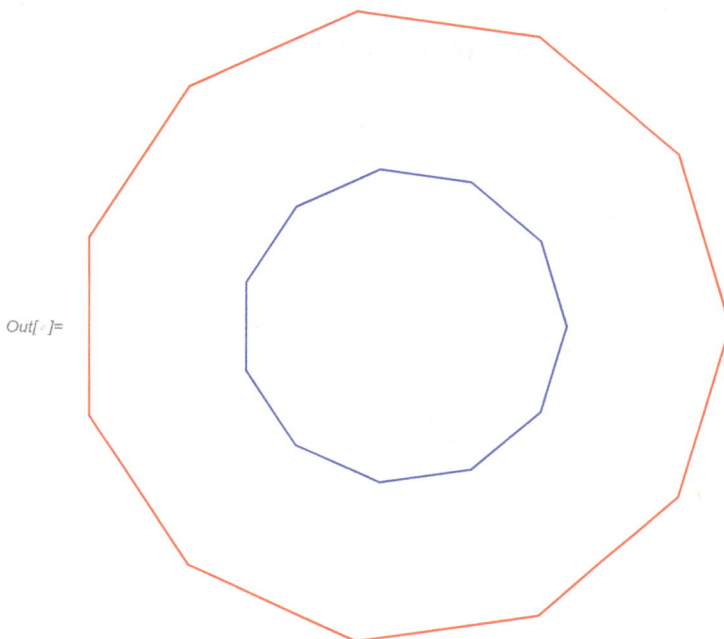

Figure 1.75

In the following dynamic picture, we can finally see the *n*th roots of unity on the unit circle.

In[·]:= Manipulate[Show[{Graphics[Circle[]],
        poly[n]}], {n, 2, 50, 1}]

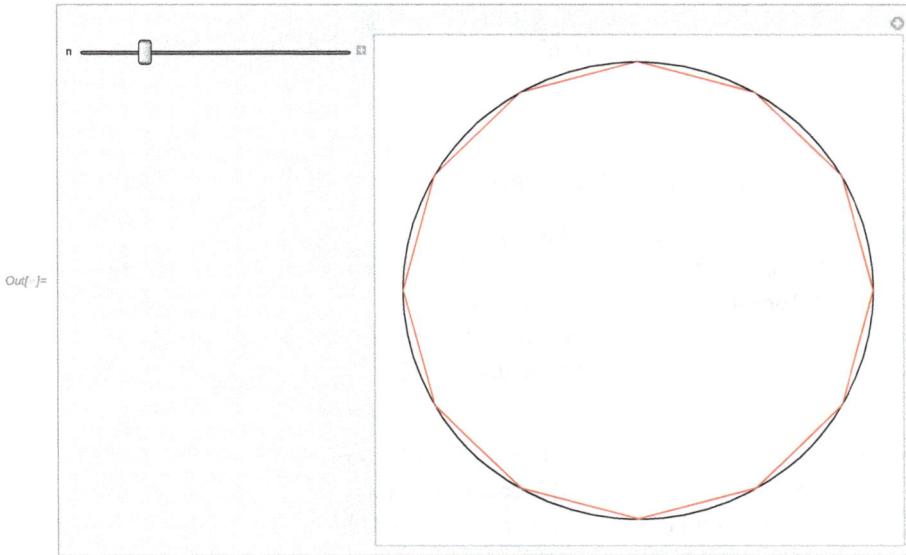

Figure 1.76

1.3 Example 2: The complex exponential map

The complex exponential map is a map $\exp : \mathbb{C} \to \mathbb{C}^*$ given by $\exp(z) = e^z = e^{\operatorname{Re} z} \cos(\operatorname{Im} z) + i \sin(\operatorname{Im} z)$. As it is known from complex analysis, this function is periodic in the complex plane. In the following dynamic picture, the image of lines and points in a rectangle around zero is visualized.

The initialization code is quite long.

```
In[·]:= labels[a_, b_, h_] := Graphics3D[{
        Style[Text["i", {0, 1.1*b, h}], Red],
        Style[Text["1", {1.1*a, 0, h}], Red]}];
```

```
In[·]:= rect[a_, b_, h_] := ParametricPlot3D[
        {u, v, h}, {u, -a, a}, {v, -b, b},
        ColorFunction -> Function[{u, v, w},
        Directive[Opacity[0.2], Brown]],
        Mesh -> False];
```

```
In[·]:= source1[a_, b_, pp_, h_] :=
        ParametricPlot3D[Table[{u, v, h},
        {u, -a, a, pp}], {v, -b, b},
        PlotStyle -> Thick, ColorFunction ->
        Function[{u, v, w}, Directive[
        Opacity[0.5], Green]]]
```

```
In[·]:= source2[a_, b_, pp_, h_] :=
        ParametricPlot3D[Table[{u, v, h},
        {v, -b, b, pp}], {u, -a, a},
        PlotStyle -> Thick,
        ColorFunction -> Function[{u, v, w},
        Directive[Opacity[0.5], Blue]]];
```

```
In[·]:= imaginary[b_, h_] := Normal[
        ParametricPlot3D[{0, v, h}, {v, -b, b},
        ColorFunction -> Function[{u, v, w},
        Directive[Red]], PlotStyle -> {Thick, Red}]]
        /. Line[x__] :> {Arrowheads[Small],
        Arrow[x]}
```

```
In[·]:= real[a_, h_] := Normal[ParametricPlot3D
        [{v, 0, h}, {v, -a, a}, ColorFunction
        -> Function[{u, v, w}, Directive[Red]],
        PlotStyle -> {Thick, Red}]] /.
        Line[x__] :> {Arrowheads[Small],
        Arrow[x]};
```

```
In[·]:= imaginaryImage[b_] := Normal[
        ParametricPlot3D[{Re[Exp[I*v]],
        Im[Exp[I*v]], 0}, {v, -b, b},
        ColorFunction -> Function[{u, v, w},
        Directive[Red]], PlotStyle ->
        {Thick, Red}]] /. Line[x__] :>
        {Arrowheads[Small], Arrow[x]}
```

```
In[·]:= realImage[a_] := Normal[
        ParametricPlot3D[{Exp[v], 0, 0},
        {v, -a, a}, ColorFunction ->
        Function[{u, v, w}, Directive[Red]],
        PlotStyle -> {Thick, Red}]] /.
        Line[x__] :> {Arrowheads[Small],
        Arrow[x]};
```

```
In[·]:= target1[a_, b_, pp_] := Normal[
        ParametricPlot3D[Table[{Re[Exp[u + I*v]],
        Im[Exp[u + I*v]], 0}, {u, -a, a, pp}],
        {v, -b, b}, PlotStyle -> {Thick,
        Directive[Opacity[0.5], Green]},
        ColorFunction -> Function[{u, v, w},
        Directive[Opacity[0.5], Green]]]] /.
        Line[x__] :> {Arrowheads[Small],
        Arrow[x]}
```

```
In[·]:= target2[a_, b_, pp_] := Normal[
        ParametricPlot3D[Table[{Re[Exp[u + I*v]],
        Im[Exp[u + I*v]], 0}, {v, -b, b, pp}],
        {u, -a, a}, PlotStyle -> {Thick,
        Directive[Opacity[0.5], Blue]},
        ColorFunction -> Function[{u, v, w},
        Directive[Opacity[0.5], Blue]]]] /.
        Line[x__] :> {Arrowheads[Small],
        Arrow[x]}
```

```
In[·]:= arrow[x_, y_, h_] := With[{z =
        x + I*y}, Graphics3D[{Dashing[0.5],
        Arrowheads[Small], Arrow[{{x, y, h},
        {Re[Exp[z]], Im[Exp[z]], 0}}]}]]
```

```
In[·]:= rect2[a_, b_] := ParametricPlot3D[{
        Re[Exp[u + I*v]], Im[Exp[u + I*v]],
        0}, {u, -a, a}, {v, -b, b},
        ColorFunction -> Function[{u, v, w},
        Directive[Opacity[0.2], Brown]],
        Mesh -> False];
```

The Manipulate code is as follows:

In[·]:= Manipulate[Show[real[a, h], imaginary[
b, h], rect[a, b, h], imaginaryImage[
b], realImage[a], labels[a, b, h],
source1[a, b, pp, h], source2[a, b, pp, h],
target1[a, b, pp], target2[a, b, pp],
rect2[a, b], If[ar, arrow[x, y, h],
Graphics3D[{Red, PointSize[0.02],
Point[{{x, y, h}, {Re[Exp[x + I*y]],
Im[Exp[x + I*y]], 0}}]}]]], PlotRange
-> All, Boxed -> False, Axes -> False,
AspectRatio -> Automatic], {{a, 1,
"size of real axis segment"}, 0.1, 2,
Appearance -> "Labeled"}, {{b, 1,
"size of imaginary axis segment"}, 0.1,
4, Appearance -> "Labeled"},
{{pp, 0.1, "grid"}, 0.01, 1,
Appearance -> "Labeled"}, {{h, 3,
"height"}, 1, 10, Appearance ->
"Labeled"}, {{x, 0.5, "real
coordinate "}, -a, a, Appearance ->
"Labeled"}, {{y, 0.5, "imaginary
coordinate "}, -b, b, Appearance
-> "Labeled"}, {{ar, True, "Show
arrow "}, {True, False}},
SaveDefinitions -> True]

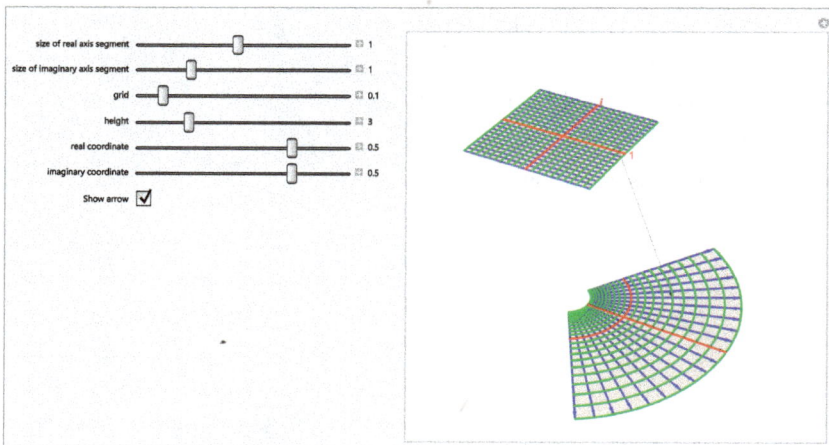

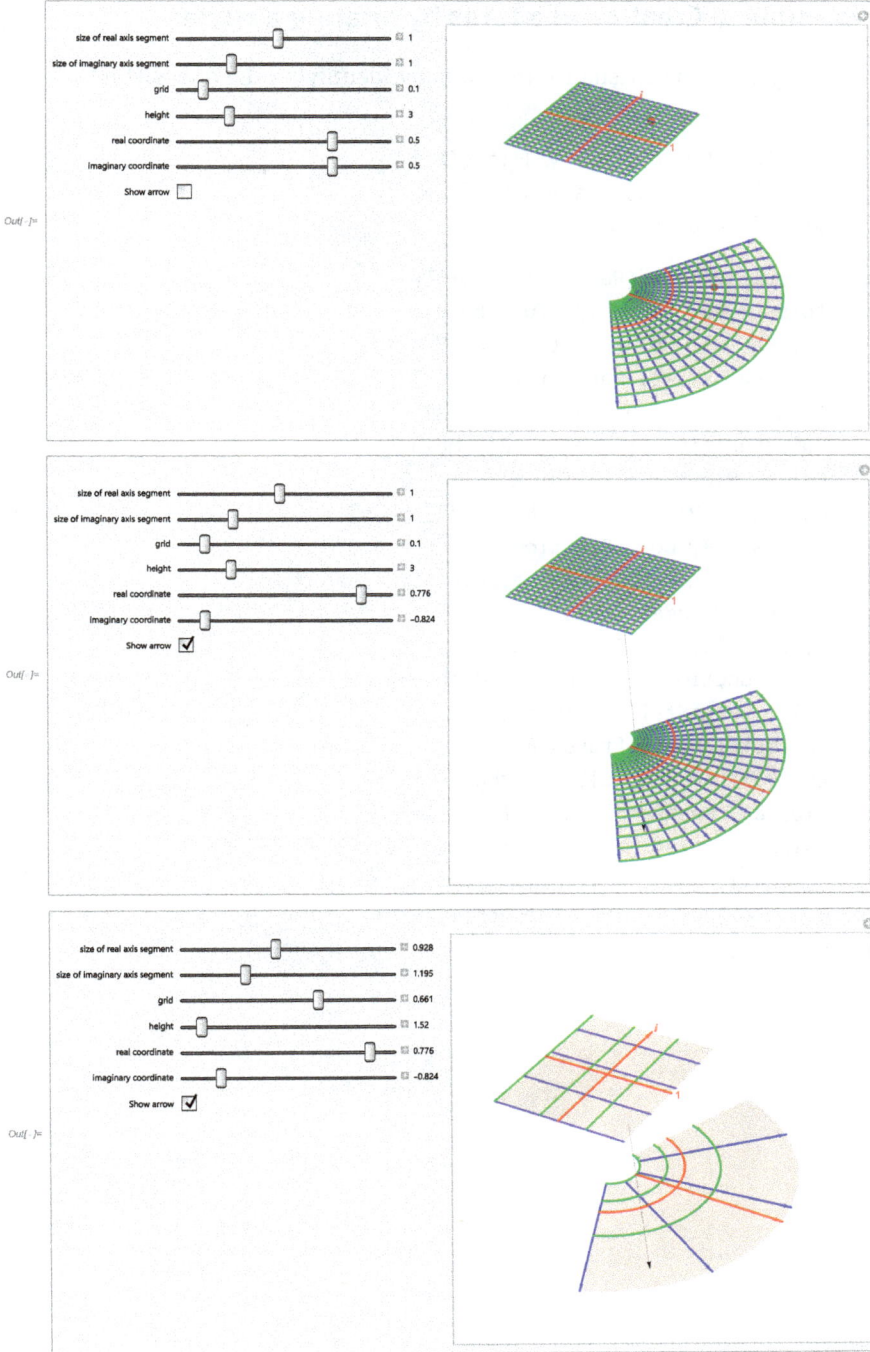

Figure 1.77

1.4 Example 3: Creation of a torus by rotating a circle

In this example we shall visualize the torus by identifying the opposite edges of a rectangle.

```
In[·]:= torus[a_, b_, c_][u_, v_] := {(a +
    b Cos[v]) Cos[u], (a + b Cos[v])
    Sin[u], c Sin[v]}
```

```
In[·]:= Manipulate[Show[{ParametricPlot3D[
    torus[8, 3, 7][u, v], {u, -b, b},
    {v, -a, a}, ColorFunction -> (Red &),
    Mesh -> False], ParametricPlot3D[
    torus[8, 3, 7][b, v], {v, -a, a},
    ColorFunction -> (Yellow &), Mesh ->
    False, PlotStyle -> Thick],
    ParametricPlot3D[torus[8, 3, 7][-b, v],
    {v, -a, a}, ColorFunction ->
    (Yellow &), Mesh -> False, PlotStyle ->
    Thick], ParametricPlot3D[torus[
    8, 3, 7][u, a], {u, -b, b},
    ColorFunction -> (Blue &), Mesh ->
    False, PlotStyle -> Thick],
    ParametricPlot3D[torus[8, 3, 7]
    [u, -a], {u, -b, b}, ColorFunction -> (
    Blue &), Mesh -> False, PlotStyle ->
    Thick]}, Boxed -> False, Axes -> False],
    {{a, 0.1, "a"}, 0.1, Pi}, {{b,
    0.1, "b"}, 0.1, Pi}, SaveDefinitions
    -> True]
```

Out[·]=

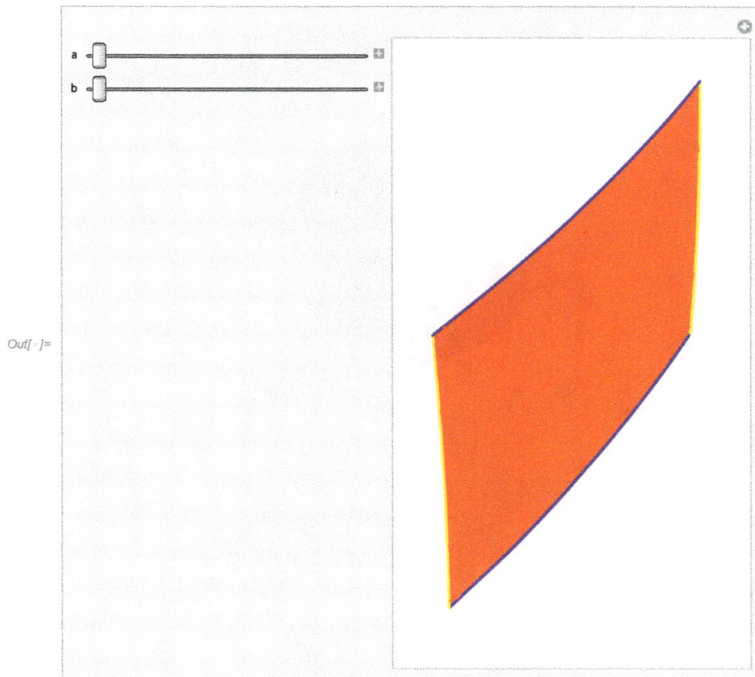

Out[·]=

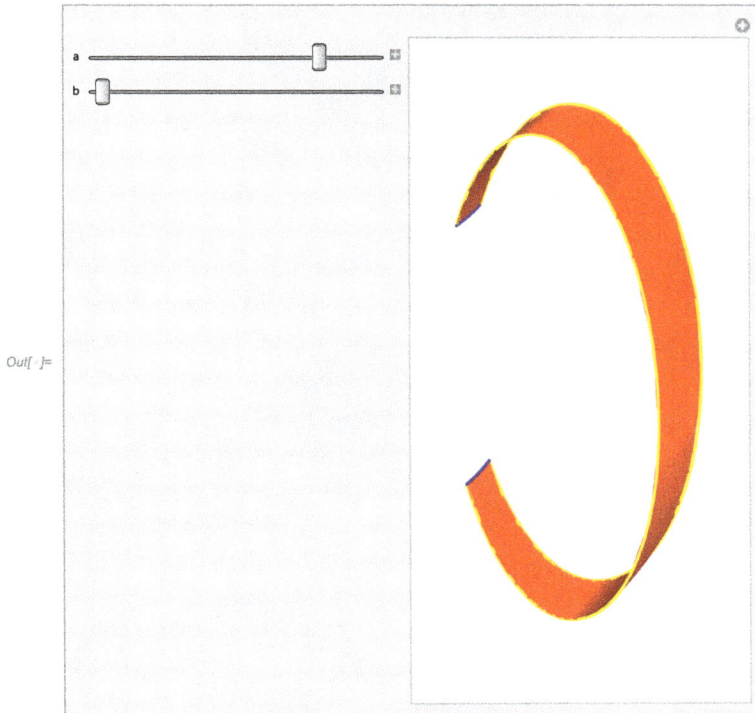

Out[ ]=

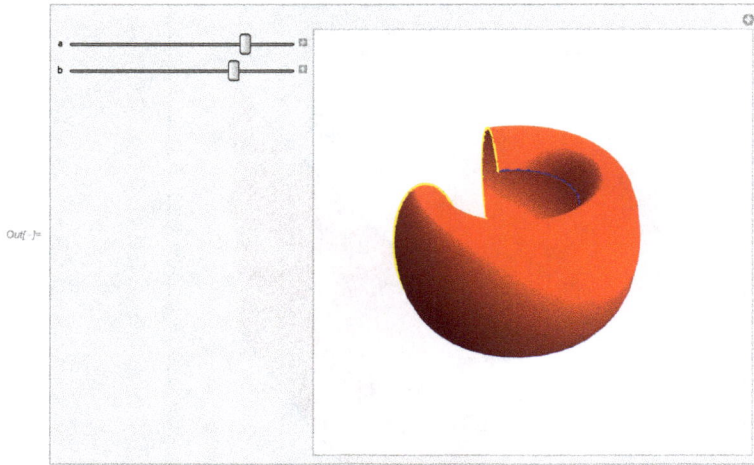

Figure 1.78

2 Regions in Mathematica®

As we have already seen in previous volumes, Mathematica® can plot regions in the plane and in space which are given by certain inequalities:

In[·]:= `RegionPlot[Inequality[1, LessEqual, x^2 + y^3,`
`Less, 2], {x, -1, 1}, {y, -1, 2}]`

Out[·]=

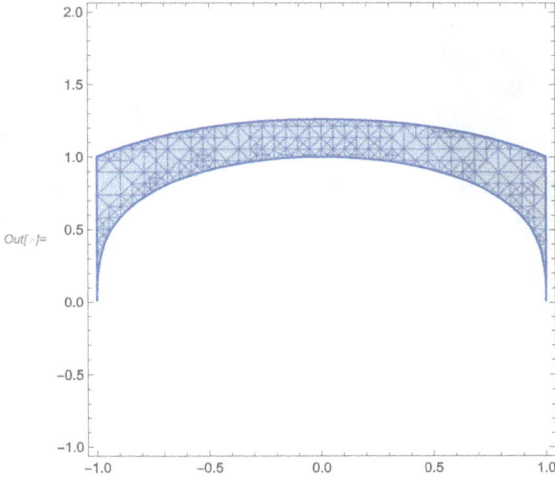

Figure 2.1

In[·]:= `RegionPlot[Inequality[1, LessEqual, x^2 + y^3,`
`Less, 2] && x < 1/2, {x, -1, 1}, {y, -1, 2}]`

Out[·]=

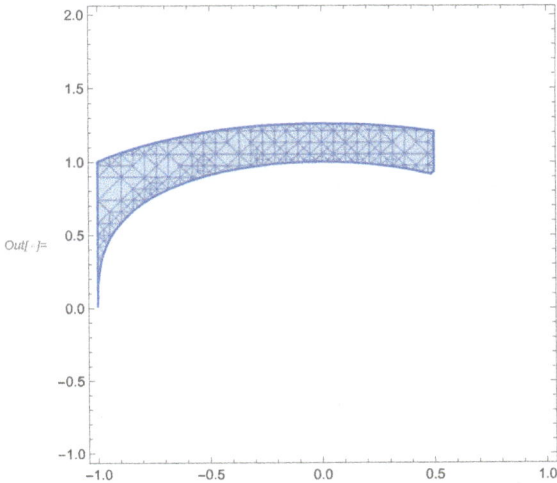

Figure 2.2

https://doi.org/10.1515/9783110774641-002

In[·]:= RegionPlot3D[x^2 + y^2 - z^2 > 4,
     {x, -2, 2}, {y, -2, 2}, {z, -2, 2}]

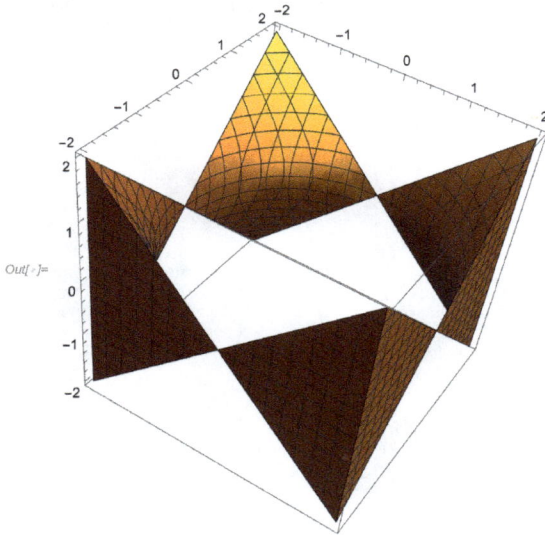

Figure 2.3

In the complex plane we can use the function ComplexRegionPlot:

In[·]:= ComplexRegionPlot[Abs[((z^3 - 1)/(z - 2))*
     Cos[z]^2] < 2, {z, 3}, PlotPoints -> 30]

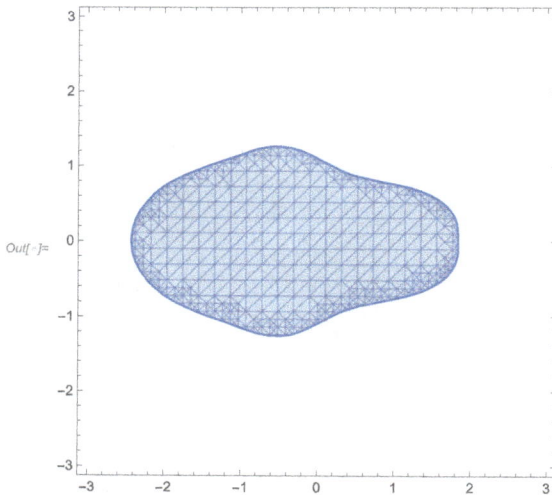

Figure 2.4

A more complicated **Mathematica**® object, which we would like to discuss in this section, is produced by Region. Although Region displays a plot, it is not a graphics object, but a new type of object, which appeared in version 11.1 and which enables a number of functions that compute geometric information to be applied. A basic example is

In[·]:= circle = Region[Circle[]]

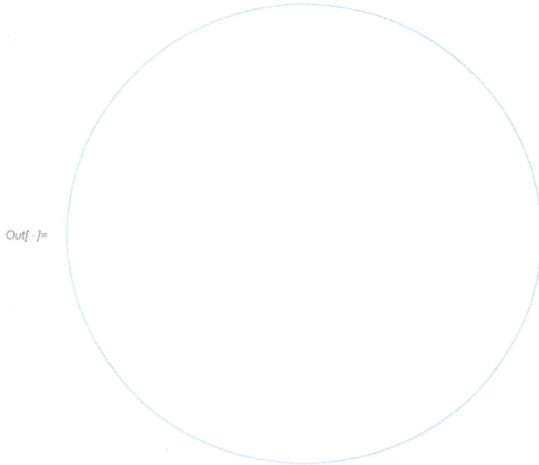

Out[·]=

Figure 2.5

This looks deceptively similar to

In[·]:= Graphics[Circle[]]

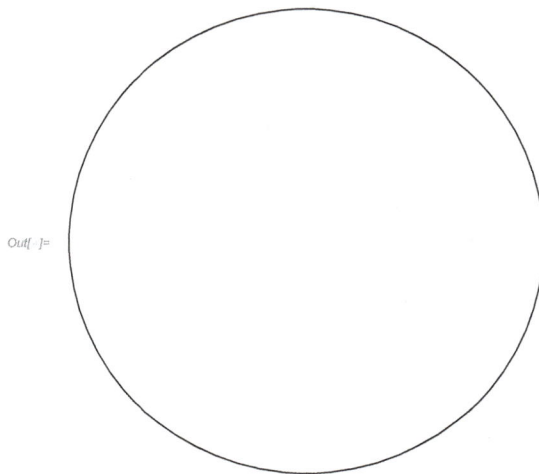

Out[·]=

Figure 2.6

but you can do more with it. For example,

In[·]:= ArcLength[circle]
Out[·]:= 2*Pi

In[·]:= RegionMeasure[circle]
Out[·]:= 2*Pi

Mathematica® knows that we draw a curve, so it is one dimensional:

In[·]:= RegionMeasure[circle, 1]
Out[·]:= 2*Pi

The two-dimensional measure is, as expected, zero:

In[·]:= RegionMeasure[circle, 2]
Out[·]:= 0

In[·]:= RegionDimension[circle]
Out[·]:= 1

In[·]:= RegionEmbeddingDimension[circle]
Out[·]:= 2

The last two outputs tell us that we are dealing with a one-dimensional object embedded in a two-dimensional Euclidean space.

Let us consider a disk

In[·]:= disk = Region[Disk[{0, 0}, 2]]

Out[·]=

Figure 2.7

In[·]:= Area[disk]
Out[·]:= 4*Pi

In[·]:= RegionMember[disk, {1/2, 1/2}]
Out[·]:= True

In[·]:= RegionDimension[disk]
Out[·]:= 2

In[·]:= RegionMeasure[disk, 2]
Out[·]:= 4*Pi

In[·]:= RegionMeasure[disk, 3]
Out[·]:= 0

We can, of course, similarly construct a sphere in three dimensions. A sphere in Mathematica® is a surface.

In[·]:= sphere = Region[Sphere[]]

Out[·]=

Figure 2.8

In[·]:= RegionMeasure[sphere]
Out[·]:= 4*Pi

We can also omit Region in the calculation below, but we need Region for visualization:

In[·]:= RegionMeasure[Sphere[]]
Out[·]:= 4*Pi

In[·]:= RegionDimension[sphere]
Out[·]:= 2

In[·]:= RegionEmbeddingDimension[sphere]
Out[·]:= 3

Note that the functions RegionPlot and RegionPlot3D crate Graphics and Graphics3D objects that are not directly related to Region. However, one can create Region objects by means of function graphs:

In[·]:= Region[ParametricRegion[{s^2, s^3},
        {{s, -1, 1}}]]

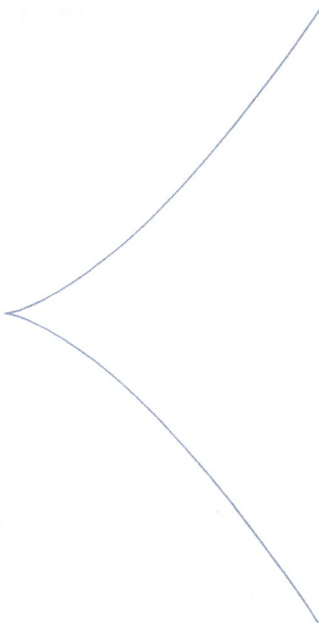

Out[·]:=

Figure 2.9

In[·]:= N[ArcLength[ParametricRegion[{s^2, s^3},
        {{s, -1, 1}}]]]
Out[·]:= 2.87942

In[·]:= Region[ParametricRegion[{{s, (1 + t)*s^2 - t},
        -1 <= s <= 1 && 0 <= t <= 1}, {s, t}]]

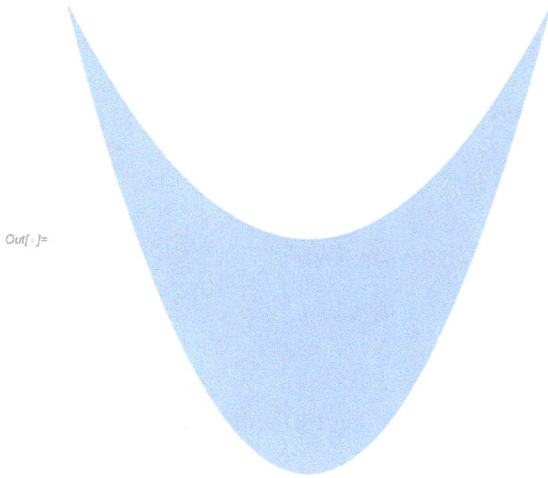

Out[ ]=

Figure 2.10

In[·]:= N[RegionMeasure[%]]
Out[·]:= 1.33333

In[·]:= N[RegionMeasure[ParametricRegion[{{s, (1 + t)*s^2 - t},
      -1 <= s <= 1 && 0 <= t <= 1}, {s, t}]]]
Out[·]:= 1.33333

In[·]:= Region[ParametricRegion[{(4 + Cos[u])*Sin[v],
      Sin[u], (4 + Cos[u])*Cos[v]},
      {{u, 0, 2*Pi}, {v, 0, 2*Pi}}]]

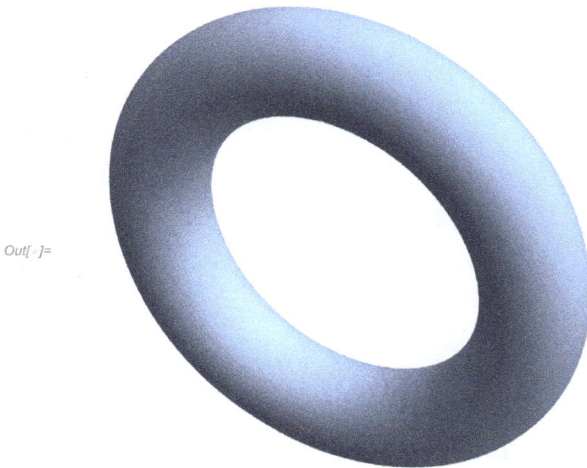

Out[ ]=

Figure 2.11

In[·]:= RegionMeasure[%]
Out[·]:= 16*Pi^2

In[·]:= RegionMeasure[ParametricRegion[{(4 + Cos[u])*Sin[v],
  Sin[u], (4 + Cos[u])*Cos[v]},
  {{u, 0, 2*Pi}, {v, 0, 2*Pi}}]]
Out[·]:= 16*Pi^2

The function ImplicitRegion can be used in the same way:

In[·]:= N[RegionMeasure[Region[ImplicitRegion[
  x^2 + y^2 <= 1, {x, y}]]]]
Out[·]:= 3.14159

How about a circle in three dimensions? A circle is not a 3D primitive, but we can use parametric equations:

In[·]:= circle3d = ParametricRegion[{Cos[t],
  Sin[t], 1}, {{t, 0, 2*Pi}}];

In[·]:= Region[circle3d, AspectRatio -> 1]

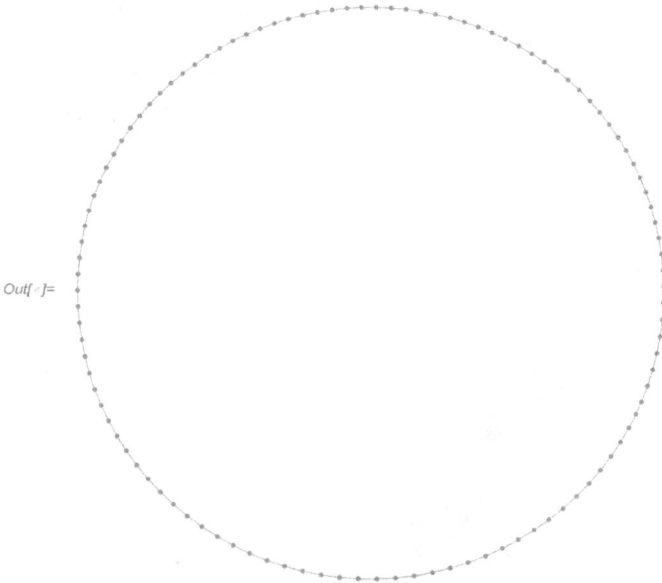

Out[·]=

Figure 2.12

In Region we can use the same options as in Graphics or Graphics3D:

In[·]:= Region[circle3d, AspectRatio -> 1,
  Boxed -> True]

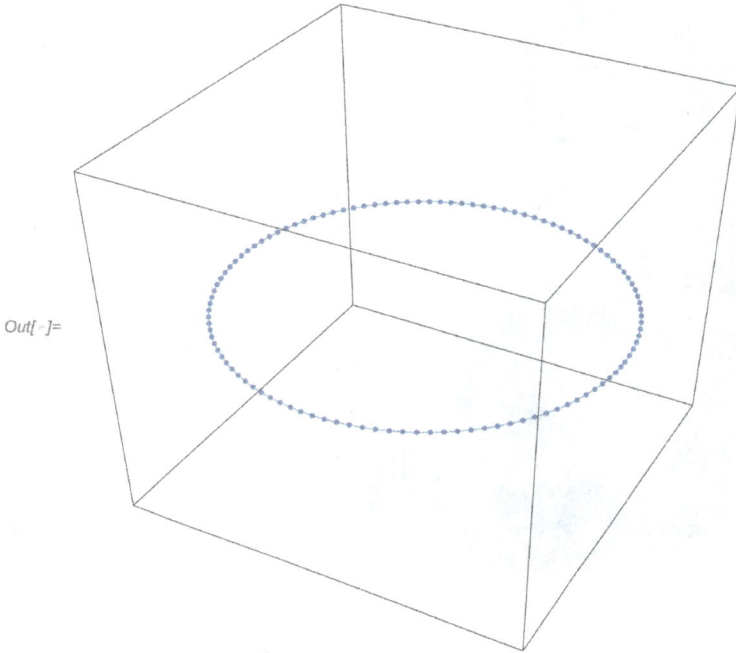

Out[ ]=

Figure 2.13

In[·]:= RegionDimension[circle3d]
Out[·]:= 1

In[·]:= RegionEmbeddingDimension[circle3d]
Out[·]:= 3

In[·]:= RegionMeasure[circle3d]
Out[·]:= 2*Pi

This, of course, is the one-dimensional measure of the three-dimensional circle.

In[·]:= RegionMeasure[circle3d, 3]
Out[·]:= 0

A ball can be defined as

In[·]:= ball = Region[Ball[3]]

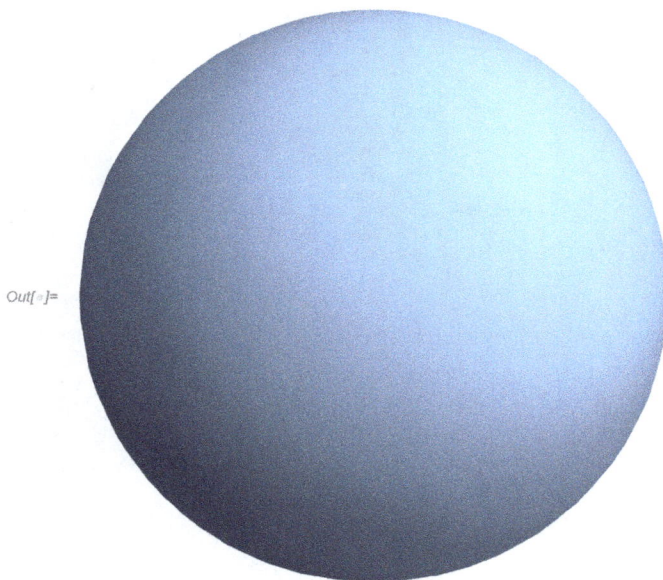

Out[ ]=

Figure 2.14

This time we have a solid ball:

In[·]:= RegionDimension[ball]
Out[·]:= 3

Note that certain geometric primitives have arbitrary dimension. In particular, Ball[n], Sphere[n] denote the unit ball and sphere in $\mathbb{R}^n$ (note that Sphere[n] has dimension $n-1$).

In[·]:= Region[Ball[4]]

Out[ ]= **Region** [◪ Embedding dimension: **4**
                          Geometric dimension: **4**]

Figure 2.15

In[·]:= RegionMeasure[Ball[4]]
Out[·]:= Pi^2/2

In[·]:= RegionBoundary[Ball[4]]
Out[·]:= Sphere[4]

In[·]:= RegionMeasure[%]
Out[·]:= 2*Pi^2

We can also create a ball in $\mathbb{R}^3$ by using ParametricRegion:

In[·]:= `ballpar = ParametricRegion[r*{Cos[u]*Cos[v],`
        `Cos[v]*Sin[u], Sin[v]}, {{u, 0, Pi}, {v, 0, 2*Pi},`
        `{r, 0, 1}}]`

Out[·]:= `ParametricRegion[{{Cos[u]*Cos[v]*r,`
        `Cos[v]*r*Sin[u], r*Sin[v]}, 0 <= u <= Pi &&`
        `0 <= v <= 2*Pi && 0 <= r <= 1}, {u, v, r}]`

The following visualization takes a lot of time:

In[·]:= `Region[ballpar]`

Out[·]=

Figure 2.16

In[·]:= `RegionDimension[ballpar]`
Out[·]:= `3`

In[·]:= `N[RegionMeasure[ballpar]]`
Out[·]:= `4.18879`

We can also use `ImplicitRegion`:

In[·]:= `ballimp = ImplicitRegion[x^2 + y^2 +`
        `z^2 <= 1, {x, y, z}]`
Out[·]:= `ImplicitRegion[x^2 + y^2 + z^2`
        `<= 1, {x, y, z}]`

The visualization this time is much faster:

In[·]:= `Region[ballimp]`

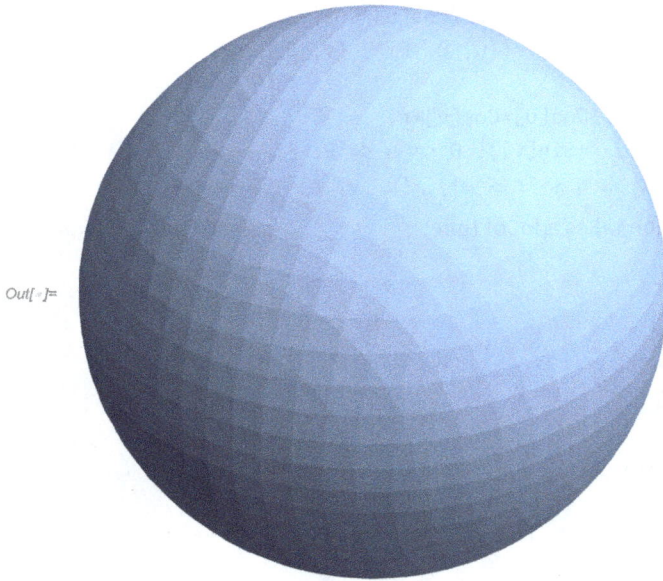

Out[ ]=

Figure 2.17

In[·]:= N[RegionMeasure[ballimp]]
Out[·]:= 4.18879

A ball in $\mathbb{R}^4$ can be defined as

In[·]:= balld4 = ImplicitRegion[x^2 + y^2 +
        z^2 + w^2 <= 1, {x, y, z, w}]
Out[·]:= ImplicitRegion[x^2 + y^2 + z^2 +
        w^2 <= 1, {x, y, z, w}]

In[·]:= RegionMeasure[balld4]
Out[·]:= Pi^2/2

In[·]:= RegionDimension[balld4]
Out[·]:= 4

Of course, a four-dimensional object cannot be represented graphically:

In[·]:= Region[balld4]

Out[ ]= **Region** [Embedding dimension: 4
                       Geometric dimension: 4]

Figure 2.18

The following function tries to find a parametric or implicit representation of a given region:

In[·]:= RegionConvert[Sphere[], "Parametric"]
Out[·]:= ParametricRegion[{{Sin[x]*Sin[y]],
   Cos[x]]*Sin[y]], Cos[y]}, 0 <= x <= 2*Pi
   && 0 <= y <= Pi}, {x, y}]

Note that the output above and below contains dots, which we omitted, underneath (formal) variables x, y, and z.

In[·]:= RegionConvert[Sphere[], "Implicit"]
Out[·]:= ImplicitRegion[x^2 + y^2 + z^2 ==
   1, {x, y, z}]

One can combine regions by means of Boolean operations:

In[·]:= Region[RegionUnion[Disk[{0, 0}, 2],
   Disk[{3, 0}, 2]]]

Out[·]=

Figure 2.19

In[·]:= Region[RegionDifference[Disk[{0, 0}, 2],
   Disk[{3, 0}, 2]]]

Out[ ]=

Figure 2.20

In[·]:= Region[RegionUnion[Sphere[], Cone[]]]

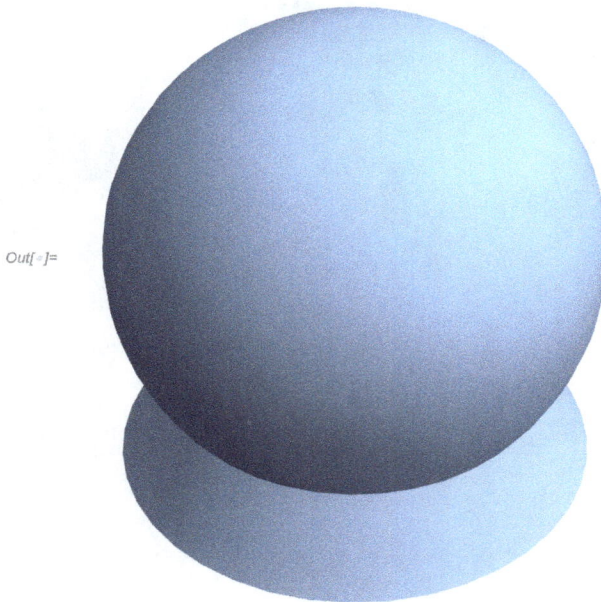

Out[ ]=

Figure 2.21

Regions can consist of pieces of different dimensions, for instance, we can combine an infinite line and a sphere:

In[·]:= infline = Region[InfiniteLine[{{3, 3, 3},
        {0, 0, 0}}]];

In[·]:= sphereline = RegionUnion[infline, Sphere[]]

Out[ ]=

Figure 2.22

In[·]:= RegionDimension[sphereline]
Out[·]:= 2

If instead of an infinite line we use an interval, we get a similar picture:

In[·]:= line = Region[Line[{{3, 3, 3}, {0, 0, 0}}]];

In[·]:= sphereline1 = RegionUnion[line, Sphere[]]

Out[ ]=

Figure 2.23

In[·]:= RegionMeasure[sphereline1, 2]
Out[·]:= 4*Pi

In[·]:= RegionMeasure[sphereline, 1]
Out[·]:= Infinity

In[·]:= RegionMeasure[sphereline1, 1]
Out[·]:= Infinity

However,

In[·]:= RegionDifference[Sphere[], infline]

Out[·]=

Figure 2.24

In[·]:= diff = RegionDifference[Sphere[], line]

Out[·]= **Region** [🖼 Embedding dimension: 3
Geometric dimension: 2]

Figure 2.25

The surface area is

In[·]:= RegionMeasure[diff, 2]
Out[·]:= 4*Pi

We can see that one point on a sphere is removed:

In[·]:= RegionMember[diff, Normalize[{1, 1, 1}]]
Out[·]:= False

In[·]:= RegionDistance[diff, Normalize[{1, 1, 1}]]
Out[·]:= 0

There are two useful functions closely related to Region, namely MeshRegion and BoundaryMeshRegion. Like Region, MeshRegion represents a region is some *n*-dimensional Euclidean space. The region is, in fact, what in topology is called a simplicial complex. Here is a one-dimensional example. We list points and indicate which of them are connected by a line:

In[·]:= meshreg = MeshRegion[{{0, 0}, {1, 0}, {2, 1/2},
    {2, -2^(-1)}}, {Line[{1, 2}], Line[{2, 3}],
    Line[{3, 4}], Line[{4, 2}]}]

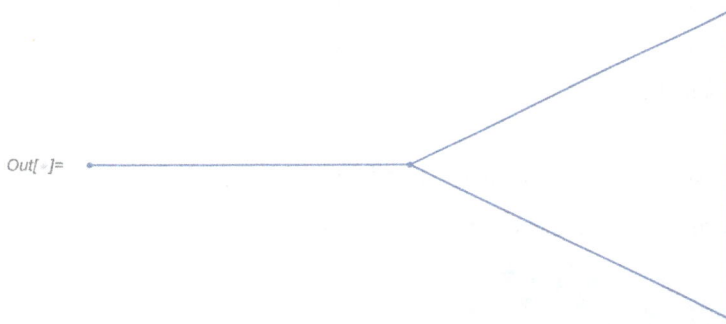

Out[ ]=

Figure 2.26

One first specifies a list of coordinates and next the objects, such as points, lines, triangles, or, more generally, simplexes determined by these coordinates. We can perform computations on a MeshRegion but only approximately.

In[·]:= RegionMeasure[meshreg]
Out[·]:= 4.23607

We can create objects with disconnected boundary with BoundaryMeshRegion:

In[·]:= BoundaryMeshRegion[{{0, 0}, {3, 0}, {3, 3},
    {0, 3}, {1, 1}, {2, 1}, {2, 2}, {1, 2}},
    Line[{1, 2, 3, 4, 1}], Line[{5, 6, 7, 8, 5}]]

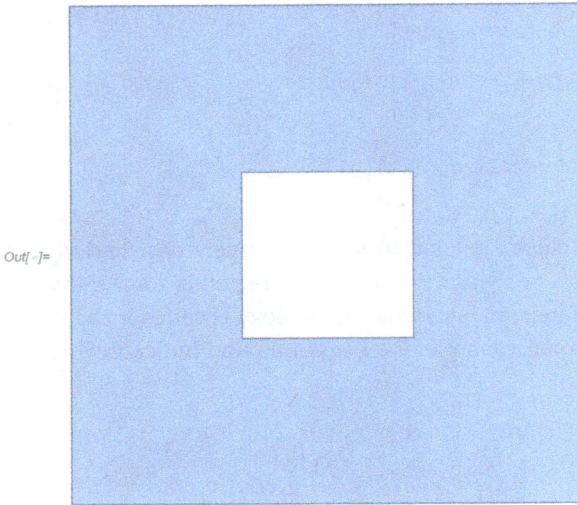

Figure 2.27

In[·]:= RegionMeasure[%]
Out[·]:= 8 .

The function DiscretizeRegion can turn a Region into a MeshRegion. For example,

In[·]:= sph = DiscretizeRegion[Sphere[]]

Figure 2.28

We can check that we obtain the correct Euler characteristic of the sphere—using the famous Euler formula (number of vertices – number of edges + number of faces):

```
In[·]:= Sum[(-1)^i*Length[MeshCells[sph, i]], {i, 0, 2}]
Out[·]:= 2
```

The unit simplex can be in arbitrary dimension:

```
In[·]:= simpl = Region[Simplex[3]]
```

Out[·]=

Figure 2.29

```
In[·]:= RegionMeasure[simpl]
Out[·]:= 1/6
```

```
In[·]:= simpl1 = Region[Simplex[4]]
```

Out[·]= **Region** [🔲 Embedding dimension: **4**
Geometric dimension: **4**]

Figure 2.30

```
In[·]:= RegionMeasure[simpl1]
Out[·]:= 1/24
```

```
In[·]:= Region[Simplex[2]]
```

Out[ ]=

Figure 2.31

In[·]:= Region[RegionBoundary[Simplex[2]]]

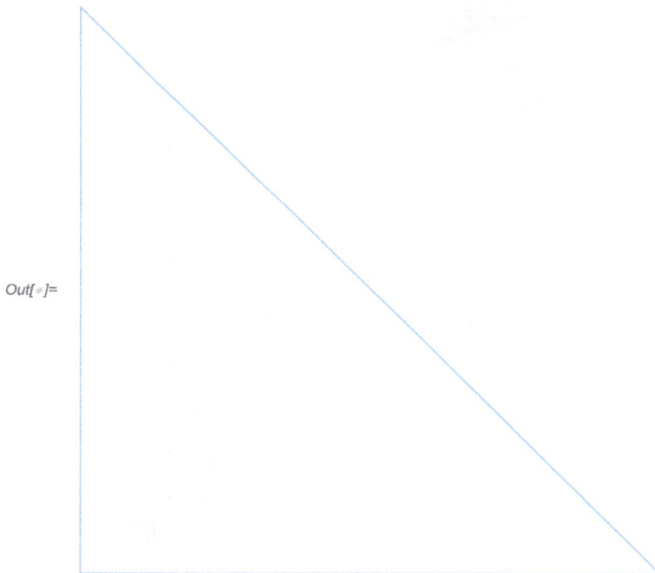

Out[ ]=

Figure 2.32

In[·]:= RegionBoundary[Simplex[4]]
Out[·]:= RegionBoundary[Simplex[4]]

In[·]:= RegionDimension[RegionBoundary[Simplex[4]]]
Out[·]:= 3

In[·]:= BoundedRegionQ[RegionBoundary[Simplex[4]]]
Out[·]:= True

In[·]:= RegionMeasure[RegionBoundary[Simplex[4]]]
Out[·]:= 1

Another way to obtain a simplex is by using BoundaryMeshRegion. However, the boundary cells must contain no self-intersections.

In[·]:= BoundaryMeshRegion[{{0, 0}, {0, 1}, {1, 0},
        {1, 1}}, Line[{1, 2, 3, 4, 1}]]
Out[·]:= BoundaryMeshRegion[{{0, 0}, {0, 1},
        {1, 0}, {1, 1}}, Line[{1, 2, 3, 4, 1}]]

In[·]:= BoundaryMeshRegion[{{0, 0}, {1, 0}, {0, 1}},
        Line[{1, 2, 3, 1}]]

Out[·]=

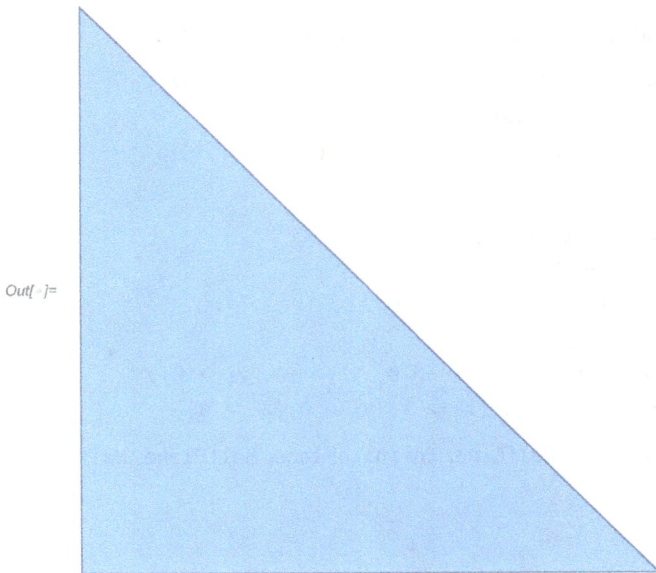

Figure 2.33

Although we can represent graphically only bounded geometric objects, **Mathematica**® allows creating unbounded ones (of course, we can represent graphically only their bounded parts). One such object, as we have already seen, is InfiniteLine. In the following example we draw a line through the point $(a/2, a^2/4)$ in the direction of the vector $(1, a)$:

```
In[·]:= Manipulate[Plot[x^2, {x, 0, a},
          Epilog -> InfiniteLine[{a/2, a^2/4}, {1, a}],
          PlotRange -> All], {a, 1, 5}]
```

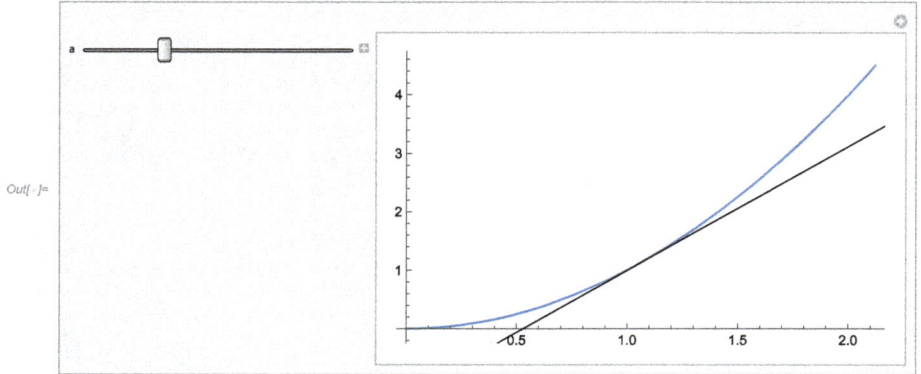

Figure 2.34

Mathematica®'s geometric objects can be used in functions such as Reduce:

```
In[·]:= N[Reduce[Element[{x, y}, InfiniteLine[{{0, 0}, {2, 2}}]]
          && Element[{x, y}, Circle[]], {x, y}]]
Out[·]:= (x == -0.7071067811865475 | x == 0.7071067811865475)|
          && y == x
```

```
In[·]:= Solve[2*x*y == z^2 && Element[{x, y, z},
          Sphere[]] && Element[{x, y, z},
          InfinitePlane[{{0, 0, 0}, {0, 1, 0},
          {1, 1, 0}}]], {x, y, z}, Reals]
Out[·]:= {{x -> -1, y -> 0, z -> 0}, {x -> 0, y -> -1, z -> 0},
          {x -> 0, y -> 1, z -> 0}, {x -> 1, y -> 0, z -> 0}}
```

Other similar infinite objects are HalfLine, InfinitePlane, HalfPlane, HalfSpace, etc.

3 Differential equations

In this chapter we will explain how differential equations can be solved with Mathematica®. First we shall revisit the notion of differentiation and then give illustrative examples of solving various differential equations and systems of differential equations and of visualizing their solutions. In this book we will mainly consider ordinary differential equations. We also recommend the reader to consult various tutorials for more information and examples, for instance, tutorial/DSolveIntroduction, tutorial/DSolveOverview, and tutorial/NDSolveOverview.

3.1 Differentiation in Mathematica® revisited

Before we start our presentation of applications of Mathematica® in differential equations and differential geometry, let us revisit the notion of differentiation in Mathematica®. We have already explained most of the following material in Volumes 1 and 2 [7, 8].

To calculate derivatives in Mathematica®, we need to distinguish between derivatives of symbolic expressions and derivatives of functions. Let us consider several examples. For more information and explanation, see Volumes 1 and 2.

In[·]:= expr = x y z
Out[·]:= x y z

The first-order derivative is calculated as

In[·]:= D[expr, x]
Out[·]:= y z

or

In[·]:= D[expr, {x}]
Out[·]:= y z

The second order derivative is calculated as

In[·]:= D[expr, {x, 2}]
Out[·]:= 0

The mixed derivative is

In[·]:= D[expr, x, y]
Out[·]:= z

or, equivalently,

In[·]:= D[expr, {x}, {y}]
Out[·]:= z

The gradient can be computed as

https://doi.org/10.1515/9783110774641-003

```
In[·]:= D[expr, {{x, y, z}}]
Out[·]:= {y z, x z, x y}
```

The built-in function Grad additionally computes the gradient in other coordinate charts or systems

```
In[·]:= Grad[expr, {x, y, z}]
Out[·]:= {y z, x z, x y}
```

```
In[·]:= CoordinateTransform["Spherical" -> "Cartesian",
          {r, u, v}]
Out[·]:= {r Cos[v] Sin[u], r Sin[u] Sin[v], r Cos[u]}
```

```
In[·]:= Simplify[Grad[expr /. Thread[{x, y, z} ->
          CoordinateTransform["Spherical" -> "Cartesian",
          {r, u, v}]], {r, u, v}, "Spherical"]]
Out[·]:= {3*r^2*Cos[u]*Cos[v]*Sin[u]^2*Sin[v],
          (1/4)*r^2*(1 + 3*Cos[2*u])*Sin[u]*Sin[2*v],
          r^2*Cos[u]*Cos[2*v]*Sin[u]}
```

The Hessian is computed by

```
In[·]:= D[expr, {{x, y, z}, 2}]
Out[·]:= {{0, z, y}, {z, 0, x}, {y, x, 0}}
```

A derivative of a function or a mapping can be computed in the following way:

```
In[·]:= cub[x_] := x^3
```

```
In[·]:= Derivative[1][cub][0]
Out[·]:= 0
```

We can also use pure functions:

```
In[·]:= cub1 = #1^3 &
Out[·]:= #1^3 &
```

```
In[·]:= Derivative[1][cub][1]
Out[·]:= 3
```

Mathematica® knows how to differentiate an inverse function:

```
In[·]:= Derivative[1][InverseFunction[f]][x]
Out[·]:= 1/Derivative[1][f][InverseFunction[f][x]]
```

and how to calculate the derivative of a composition of functions:

```
In[·]:= Derivative[1][g @* h][x]
Out[·]:= Derivative[1][g][h[x]]*Derivative[1][h][x]
```

For functions of several variables, we have

In[·]:= quadr[x_, y_] := x^2 + y^2

In[·]:= Derivative[1, 0][quadr][x, y]
Out[·]:= 2*x

In[·]:= Derivative[0, 1][quadr]
Out[·]:= 2*#2 &

In[·]:= Module[{u, v}, D[quadr[u, v], v]
        /. {u -> 1, v -> 2}]
Out[·]:= 4

In[·]:= Derivative[0, 1][quadr][1, 2]
Out[·]:= 4

In[·]:= Derivative[1, 1][Function[{x, y}, x*y^2]]
Out[·]:= Function[{x, y}, 2*y]

In[·]:= Derivative[1, 1][#1*#2^3 &][u, v]
Out[·]:= 3*v^2

In[·]:= quadr1[x_, y_] := {x^2 - y^2 + 2*x,
        x^2 - 3*x^2 + 1}

In[·]:= D[quadr1[x, y], {{x, y}}]
Out[·]:= {{2 + 2*x, -2*y}, {-4*x, 0}}

3.2 Built-in functions to solve differential equations and systems

In this section we consider built-in functions which allow us to solve simple scalar differential equations and systems of differential equations. Note that in general there are very few methods to solve nonlinear differential equations exactly and more analysis is needed to extract information about the behavior of solutions.

The basic functions DSolve and NDSolve have a lot of built-in methods. Given a differential equation (with or without initial conditions), we can try to solve it by using DSolve. For instance,

In[·]:= DSolve[Derivative[1][y][x] - y[x] == 0, y[x], x]
Out[·]:= {{y[x] -> E^x*C[1]}}

In[·]:= DSolve[{Derivative[1][y][x] - y[x] == 0,
        y[0] == 1}, y[x], x]
Out[·]:= {{y[x] -> E^x}}

If we use y instead of y[x], we will obtain a pure function:

```
In[·]:= sol = DSolve[{Derivative[1][y][x] - y[x] == 0,
        y[0] == 1}, y, x]
Out[·]:= {{y -> Function[{x}, E^x]}}
```

```
In[·]:= f = y /. sol[[1]]
Out[·]:= Function[{x}, E^x]
```

We can plot the solution using

```
In[·]:= Plot[f[x], {x, 0, 2}, PlotRange -> All]
```

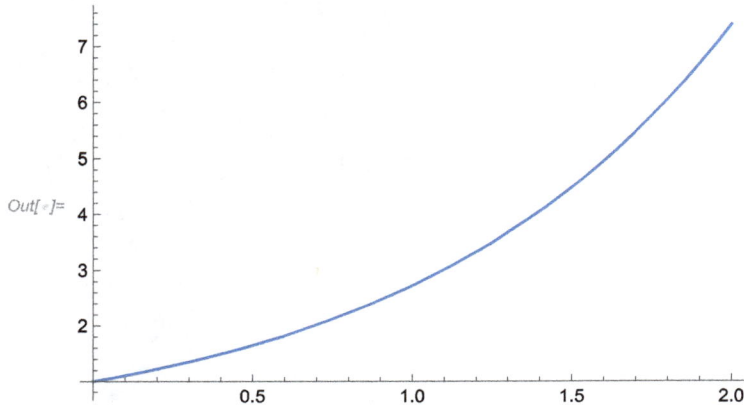

Figure 3.1

Most of differential equations cannot be solved exactly. In such cases we can use numerical methods and the built-in functions NDSolve or ParametricNDSolve (if we have parameters in the equation)

```
In[·]:= y[x] /. NDSolve[{Derivative[2][y][x] + 25*y[x] == 0,
        y[0] == 1, Derivative[1][y][0] == 5}, y[x],
        {x, 0, 2}][[1]][[1]]
```

Out[·]= InterpolatingFunction[⊞ ⋀ Domain: {{0., 2.}} Output: **scalar**] [x]

Figure 3.2

The following function

```
In[·]:= pl=Plot[%, {x, 0, 2}]
```

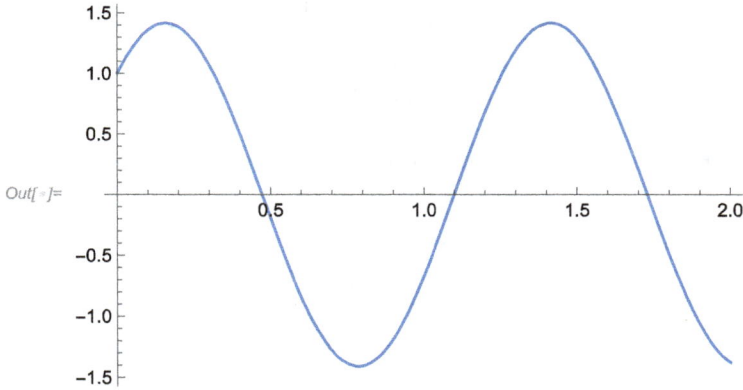

Figure 3.3

will give exactly the same plot as the exact solution. Note that if we compute a solution over the region $0 < x < 1$ but plot it over the bigger region $0 < x < 2$, Mathematica® will produce a different plot

```
In[·]:= f1 = y /. NDSolve[{Derivative[2][y][x] +
        25*y[x] == 0, y[0] == 1,
        Derivative[1][y][0] == 5}, y, {x, 0, 1}][[1]]
```

Out[·]= InterpolatingFunction[⊞ Domain: {{0., 1.}} Output: scalar]

Figure 3.4

```
In[·]:= Plot[f1[x], {x, 0, 2}]
```

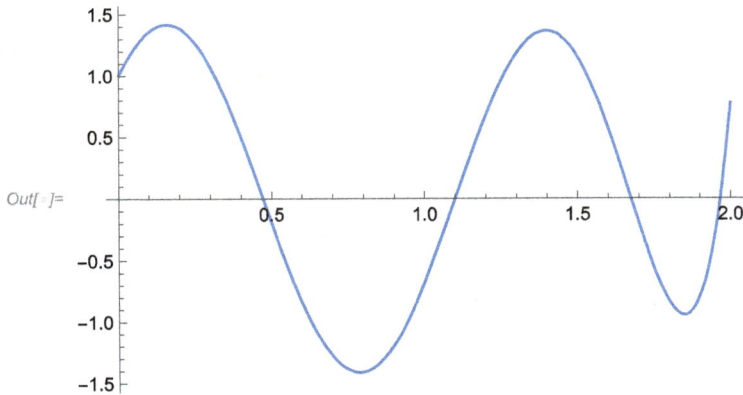

Figure 3.5

This is because extrapolation methods were used to extend the solution and they are not reliable. Therefore, when using NDSolve we need to indicate precisely the domain for the independent variable for which **Mathematica**® should give the result and then use it in other computations.

Another useful function is ParametricNDSolve. Let us introduce a parameter into the equation:

In[·]:= f2 = y /. ParametricNDSolve[{Derivative[2][y][x]
    + a*y[x] == 0, y[0] == 1, Derivative[1][y][0]
    == 5}, y, {x, 0, 2}, {a}]

Out[·]= ParametricFunction[⊞ 〰 Expression: y Parameters: {a}]

Figure 3.6

We can compare the values at $x = 1/2$ for the functions f1 and f2 with parameter $a = 25$ and see that they are exactly the same:

In[·]:= f1[1/2]
Out[·]:= -0.202671

In[·]:= f2[25][1/2]
Out[·]:= -0.202671

In[·]:= Plot[f2[25][x], {x, 0, 2}]

The plot of the solution for $a = 25$ is exactly the same as pl above, so we omit the picture.

We can use Manipulate to demonstrate the behavior of solutions for different values of the parameter a:

In[·]:= Manipulate[Plot[f2[a][x], {x, 0, 2}], {a, -25, 25}]

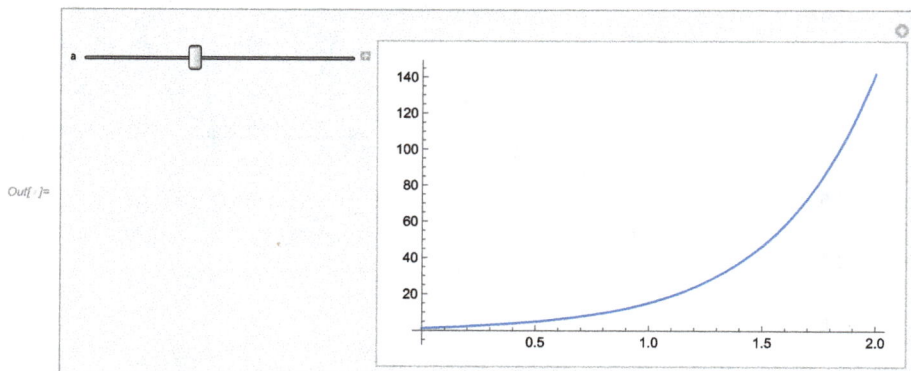

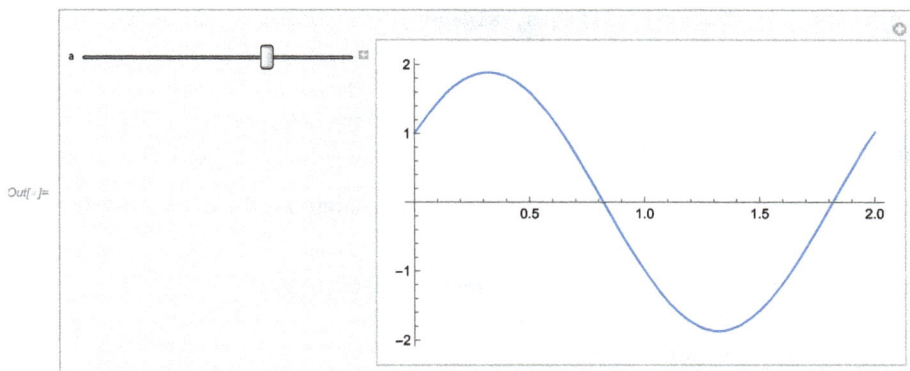

Figure 3.7

Of course, in this case **Mathematica**® can solve equation exactly, but for more compli-cated equations ParametricNDSolve may prove useful.

```
In[·]:= Assuming[a < 0, DSolve[{Derivative[2][y][x] +
        a*y[x] == 0, y[0] == 1, Derivative[1][y][0]
        == 5}, y[x], x]]
Out[·]:= {{y[x] -> (Sqrt[-a]*Cosh[Sqrt[-a]*x] +
        5*Sinh[Sqrt[-a]*x])/Sqrt[-a]}}

In[·]:= Assuming[a > 0, DSolve[{Derivative[2][y][x] +
        a*y[x] == 0, y[0] == 1, Derivative[1][y][0]
        == 5}, y[x], x]]
Out[·]:= {{y[x] -> (Sqrt[a]*Cos[Sqrt[a]*x] +
        5*Sin[Sqrt[a]*x])/Sqrt[a]}}
```

Mathematica® can solve differential systems of equations combined with algebraic constraints. For instance, in the following example **Mathematica**® cannot solve ex-actly the following system of equations but can solve it numerically:

```
In[·]:= diffeqns = {Derivative[2][x][t] == a[t]*x[t]*y[t],
        Derivative[2][y][t] == a[t]*y[t] - x[t]};
        algeqns = {2*x[t]^2 + 3*y[t]^2 == 10};
        initconds = {x[0] == 1, Derivative[1][y][0] == 1};

In[·]:= DSolve[{diffeqns, algeqns, initconds}, {x, y, a}, t];
```

(We omit the output as it is the list of the same equations.)

In[·]:= sol1a = NDSolve[{diffeqns, algeqns, initconds},
    {x, y, a}, {t, 0, 4}];

     ··· NDSolve: The DAE solver failed at $t = 0$.`. The solver is intended for index
1 DAE systems and structural analysis indicates that the DAE index is 3. The option
Method->{``IndexReduction''->Automatic} may be used to reduce the index of the system.

In this case Mathematica® suggests to use a certain method. If we use it, Mathematica® gives us an answer:

In[·]:= sol1 = NDSolve[{diffeqns, algeqns, initconds},
    {x, y, a}, {t, 0, 4}, Method ->
    {"IndexReduction" -> Automatic}]

Out[·]:= {{x → InterpolatingFunction[⊞ \/ Domain: {{0., 4.}} Output: scalar],

    y → InterpolatingFunction[⊞ \/ Domain: {{0., 4.}} Output: scalar], a → InterpolatingFunction[⊞ \/\/ Domain: {{0., 4.}} Output: scalar]}}

Figure 3.8

In[·]:= f3 = x /. sol1[[1]]

Out[·]:= InterpolatingFunction[⊞ \/ Domain: {{0., 4.}} Output: scalar]

Figure 3.9

In[·]:= Plot[f3[t], {t, 0, 4}]

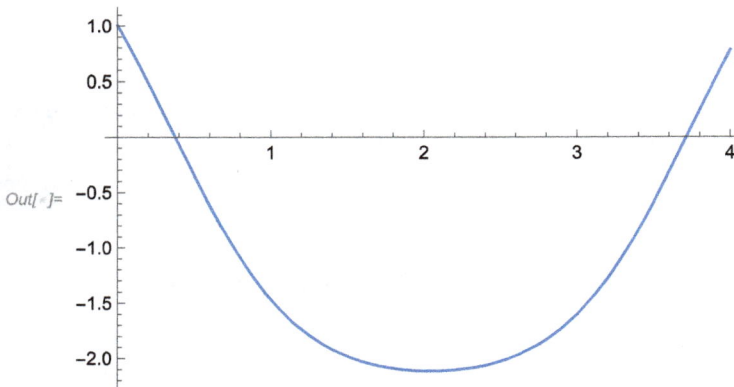

Figure 3.10

In[·]:= Plot[Evaluate[{x[t], y[t]} /. sol1], {t, 0, 4}]

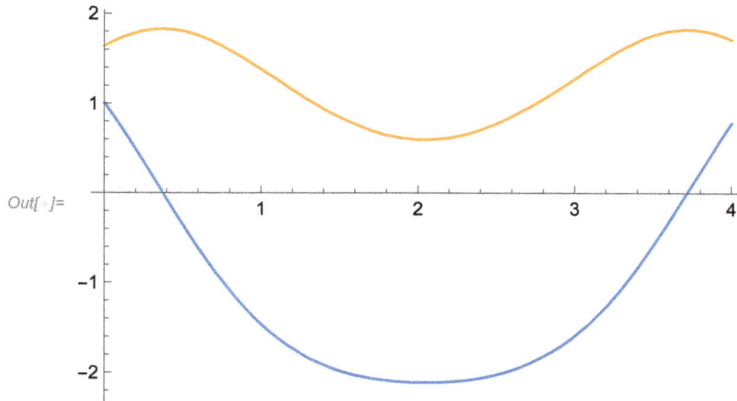

Figure 3.11

A lot of physical phenomena can be described by differential equations. For instance, the following equations describe the motion of the double pendulum (see also example/DoublePendulum):

```
In[·]:= diffeqns1 = {Derivative[2][x1][t] ==
        a1[t]*x1[t] - a2[t]*(x2[t] - x1[t]),
        Derivative[2][y1][t] == a1[t]*y1[t] -
        a2[t]*(y2[t] - y1[t]) - 9, Derivative[2][x2][t]
        == a2[t]*(x2[t] - x1[t]), Derivative[2][y2][t] ==
        a2[t]*(y2[t] - y1[t])}; algeqns1 = {x1[t]^2 + y1[t]^2
        == 1, (x2[t] - x1[t])^2 + (y2[t] - y1[t])^2 == 1};
        initconds1 = {x1[0] == 1, Derivative[1][y1][0] ==
        1, x2[0] == 2, Derivative[1][y2][0] == 1};
```

There is the whole section in **Mathematica**® documentation center which can be found by typing tutorial/NDSolveDAE devoted to such systems. One can read about Projection method for NDSolve used below in [28].

```
In[·]:= sol4 = NDSolve[{diffeqns1, algeqns1,
        initconds1}, {x1, y1, x2, y2, a1, a2},
        {t, 0, 15}, Method -> {"IndexReduction" ->
        {True, "ConstraintMethod" -> "Projection"}}];
```

```
In[·]:= Animate[Graphics[{{Thickness[0.01], Orange,
        Line[{{0, 0}, {x1[t], y1[t]}}]}, {Thickness[0.01],
        Green, Line[{{x1[t], y1[t]}, {x2[t], y2[t]}}]}}
        /. sol4, PlotRange -> {{-2.1, 2.1}, {0.6, -2.1}},
        Frame -> True], {t, 0, 15}, SaveDefinitions ->
        True, AnimationRunning -> False]
```

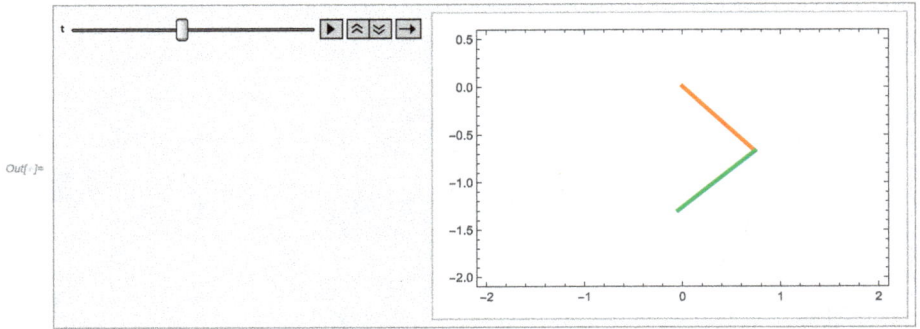

Figure 3.12

Another useful function is `ParametricNDSolveValue` which gives the value of an expression with functions determined by a numerical solution (in a certain range) to the ordinary differential equations with parameters. For instance, in the following we obtain a function of the parameter *a* that gives the value of the function *y* at *t* = 5:

```
In[·]:= sol2 = ParametricNDSolveValue[{
         Derivative[1][y][t] + a*y[t] == 0, y[0] ==
         1}, y[5], {t, 0, 10}, {a}]
```

Out[·]= ParametricFunction[⊞ 〽 Expression: y[5] Parameters: {a}]

Figure 3.13

We can now plot the value as a function of the parameter *a*:

```
In[·]:= Plot[sol2[a], {a, 0, 2}]
```

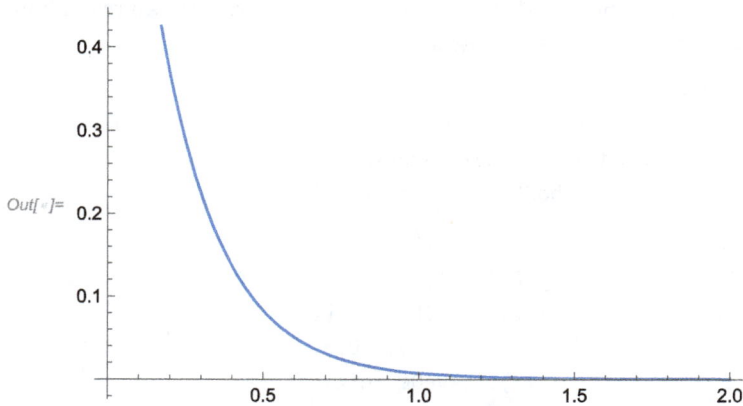

Figure 3.14

We can have more parameters and plot the solution for various values of these parameters:

```
In[·]:= sol3 = ParametricNDSolveValue[{
        Derivative[2][y][x] + a*Derivative[1][y][x] +
        b*y[x] == 0, y[0] == 1, Derivative[1][y][0]
        == 1}, y, {x, 0, 10}, {a, b}]
```

Out[·]= ParametricFunction[🞣 〽 Expression: y Parameters: {a, b}]

Figure 3.15

```
In[·]:= Manipulate[Plot[Evaluate[sol3[a, b][x]],
        {x, 0, 10}, PlotRange -> All], {x, 0, 10},
        PlotRange -> All], {{a, 3, "a"}, -10, 10,
        Appearance -> "Labeled"}, {{b, 2, "b"}, -10, 10,
        Appearance -> "Labeled"}]
```

Out[·]=

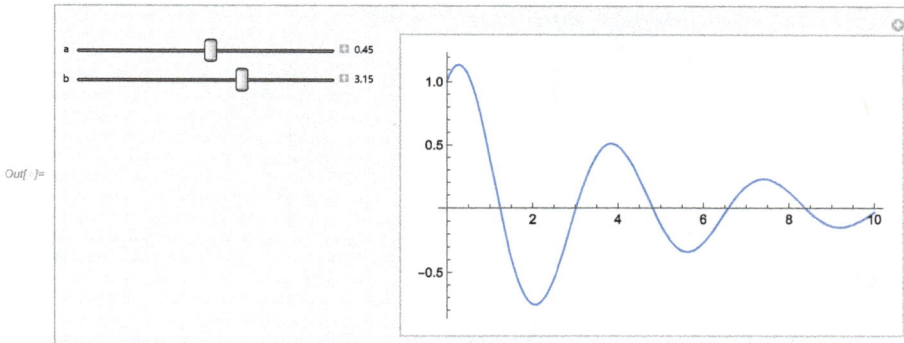

Figure 3.16

Again, in this case **Mathematica**® can solve this second-order differential equation exactly with arbitrary parameters (we omit the result as it is quite long):

```
In[·]:= DSolve[{Derivative[2][y][x] + a*Derivative[1][y][x] +
        b*y[x] == 0, y[0] == 1, Derivative[1][y][0]
        == 1}, y, x];
```

We can choose the point for our initial conditions on the graph and vary parameters to understand better the behavior of solutions:

```
In[·]:= Manipulate[With[{sol = NDSolve[{
        Derivative[2][y][t] - y[t] == a*Sin[b*t + c],
        y[0] == pt[[1]], Derivative[1][y][0] == pt[[2]]},
        y, {t, 0, 50}]}, ParametricPlot[{y[t],
        Derivative[1][y][t]} /. sol, {t, 0, t1},
        PlotRange -> 5]], {a, -3, 3}, {b, -Pi, Pi},
        {c, 1, 5}, {t1, 10, 50}, {{pt, {0, 0}}, Locator},
         SaveDefinitions -> True]
```

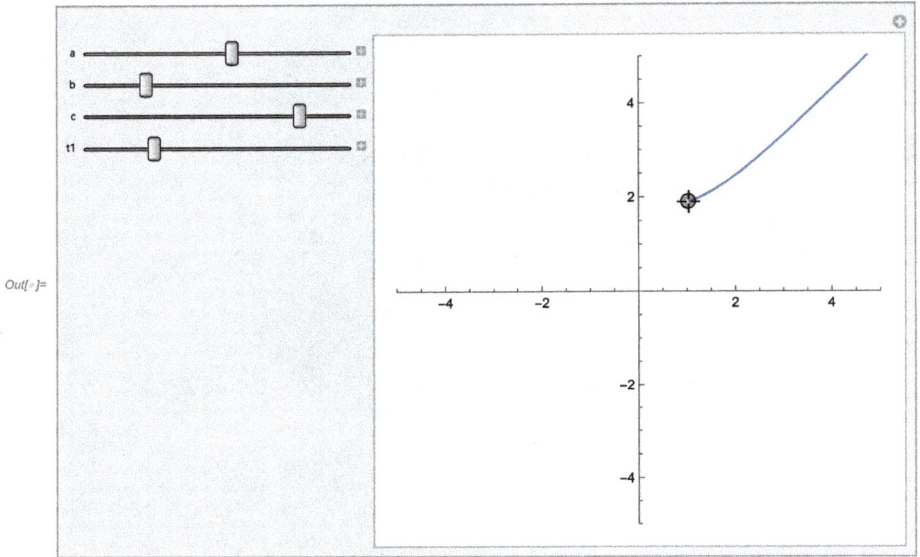

Figure 3.17

In version 11.3 the function AsymptoticDSolveValue was introduced. It computes an asymptotic approximation to the solution of a differential equation. There are other functions like Asymptotic, AsymptoticIntegrate, AsymptoticSolve, and AsymptoticRSolveValue which compute certain asymptotic expansions of functions, integrals, transcendental equations, and difference equations. Let us consider a few examples.

```
In[·]:= Asymptotic[Sin[x], x -> 0, SeriesTermGoal -> 10]
Out[·]:= x - x^3/6 + x^5/120 - x^7/5040 + x^9/362880
```

```
In[·]:= Asymptotic[Sin[x], x -> 1]
Out[·]:= Sin[1]
```

```
In[·]:= AsymptoticIntegrate[E^(t*x), {t, 0, 1}, {x, 0, 3}]
Out[·]:= 1 + x/2 + x^2/6 + x^3/24
```

In this case we could have achieved the same result by

In[·]:= `Normal[Series[Integrate[E^(t*x), {t, 0, 1}],`
`{x, 0, 3}]]`
Out[·]:= `1 + x/2 + x^2/6 + x^3/24`

In[·]:= `AsymptoticSolve[E^y - Cos[x*y] + x^10 == 0,`
`{y, 0}, {x, 0, 30}]`
Out[·]:= `{{y -> -x^10 - x^20/2 - x^22/2 - x^30/3}}`

In[·]:= `AsymptoticRSolveValue[y[n + 1] == (n + 1)*y[n],`
`y[n], {n, Infinity, 1}]`
Out[·]:= `((1 + 1/(12*n))*n^(1/2 + n)*C[1])/E^n`

For differential equations we can find expansions of solutions in a neighborhood of a given point ($x = 0$ and $x = 1$ in the examples below):

In[·]:= `AsymptoticDSolveValue[{Derivative[2][y][x] +`
`y[x] == 0, y[0] == 1, Derivative[1][y][0] ==`
`0}, y[x], {x, 0, 4}]`
Out[·]:= `1 - x^2/2 + x^4/24`

In[·]:= `AsymptoticDSolveValue[{Derivative[2][y][x] +`
`y[x] == 0, y[0] == 1, Derivative[1][y][0]`
`== 0}, y[x], {x, 1, 2}]`
Out[·]:= `Cos[1] - (1/2)*(-1 + x)^2*Cos[1] - (-1 + x)*Sin[1]`

A differential equation may also contain a parameter, and we can search for the expansion of a solution in this parameter:

In[·]:= `AsymptoticDSolveValue[{p*Derivative[2][y][x] +`
`Derivative[1][y][x] + x*y[x] == 0, y[0] == 0,`
`y[1] == 1/2}, y[x], x, {p, 0, 1}]`
Out[·]:= `(-(1/2))*E^(1/2 - x/p) + (1/2)*E^(1/2 - x^2/2)`

In this case **Mathematica**® returns the result very fast. If we first try to solve the equation with initial values explicitly, we get a complicated expression given in terms of the Airy functions, with arguments involving the parameter p, and so that a series expansion in p becomes complicated (we omit the formulas but the interested reader may try to evaluate the following cells):

In[·]:= `Simplify[y[x] /. DSolve[{p*Derivative[2][y][x] +`
`Derivative[1][y][x] + x*y[x] == 0, y[0] == 0,`
`y[1] == 1/2}, y[x], x][[1]]];`

In[·]:= `Series[%, {p, 0, 1}];`

Sometimes the same output can be achieved by combining two functions instead of one function. For instance, the following cells return exactly the same output:

In[·]:= AsymptoticDSolveValue[{Derivative[2][y][x] -
    Derivative[1][y][x] + x^10*y[x] == 1, y[0] == 1,
    Derivative[1][y][0] == 2}, y[x], {x, 0, 4}]
Out[·]:= 1 + 2*x + (3*x^2)/2 + x^3/2 + x^4/8

In[·]:= Asymptotic[DSolveValue[{Derivative[2][y][x]
    - Derivative[1][y][x] + x^10*y[x] == 1, y[0] == 1,
    Derivative[1][y][0] == 2}, y[x], x], {x, 0, 4}]
Out[·]:= 1 + 2*x + (3*x^2)/2 + x^3/2 + x^4/8

Mathematica® actually solves this equation using `DifferentialRoot` with no explicit branch cuts, and we can expand the solution at $x = 0$ to get the same expansion:

In[·]:= DSolve[{Derivative[2][y][x] - Derivative[1][y][x]
    + x^10*y[x] == 1, y[0] == 1, Derivative[1][y][0]
    == 2}, y[x], x]

Out[·]= $\left\{\left\{y[x] \rightarrow \boxed{\partial\ y''[x]\ [x]}\right\}\right\}$

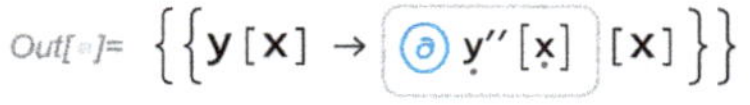

Figure 3.18

In[·]:= Series[y[x] /. %[[1]], {x, 0, 4}]
Out[·]:= SeriesData[x, 0, {1, 2, 3/2, 1/2, 1/8}, 0, 5, 1]

In version 12.3 new features for solving ODEs were introduced. A new functionality to solve linear systems of ODEs with rational function coefficients was developed. Moreover, Mathematica® now solves a linear ODE with q-rational function coefficients. Mathematica® can also solve delay equations:

In[·]:= soldel = DSolve[{Derivative[1][x][t] ==
    Sign[x[t - 2]] - x[t - 2], x[t /; t <= 0]
    == 0.2}, x, {t, 0, 20}];

In[·]:= ParametricPlot[Evaluate[{x[t - 2],
    x[t]} /. soldel], {t, 2, 20}]

Out[ ]=

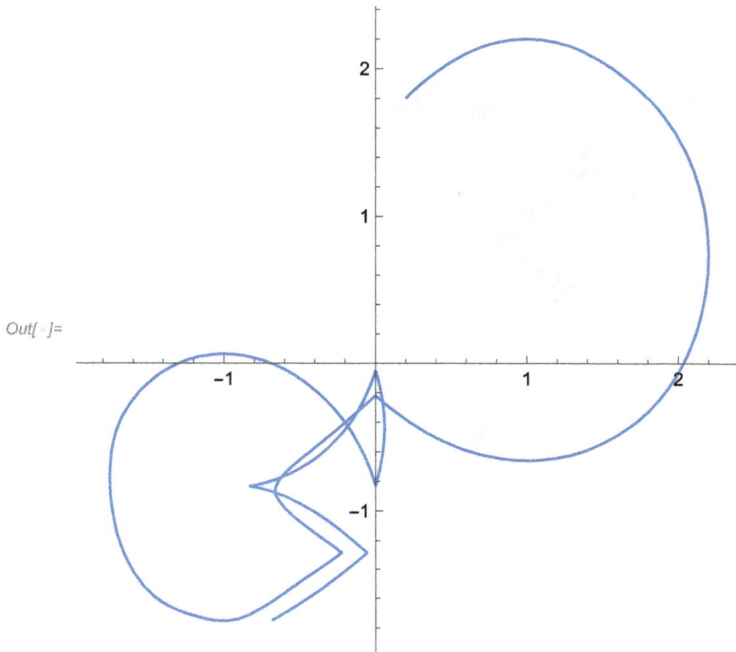

Figure 3.19

Finally, we would like to remark that we can use regions for initial conditions, especially when solving partial differential equations, for instance,

In[·]:= NDSolve[{Laplacian[u[x, y], {x, y}] == 0,
        DirichletCondition[u[x, y] == Sin[x*y],
        True]}, u, Element[{x, y}, Disk[]]]

Out[ ]= {{u → InterpolatingFunction[⊞ 〰 Domain: {{-1., 1.}, {-1., 1.}} Output: scalar]}}

Figure 3.20

In[·]:= Plot3D[Evaluate[u[x, y] /. %],
        Element[{x, y}, Disk[]]]

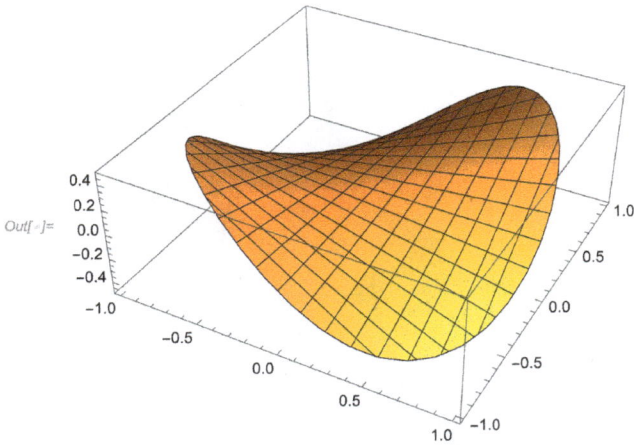

Figure 3.21

In[·]:= NDSolve[{Laplacian[u[x, y], {x, y}] == 0,
      DirichletCondition[u[x, y] == Sin[x*y],
      y > 0]}, u, Element[{x, y}, Disk[]]]

Out[·]= {{u → **InterpolatingFunction**[⊞ 〰 Domain: {{-1., 1.}, {-1., 1.}}
                                            Output: **scalar**]}}

Figure 3.22

In[·]:= Plot3D[Evaluate[u[x, y] /. %],
      Element[{x, y}, Disk[]]]

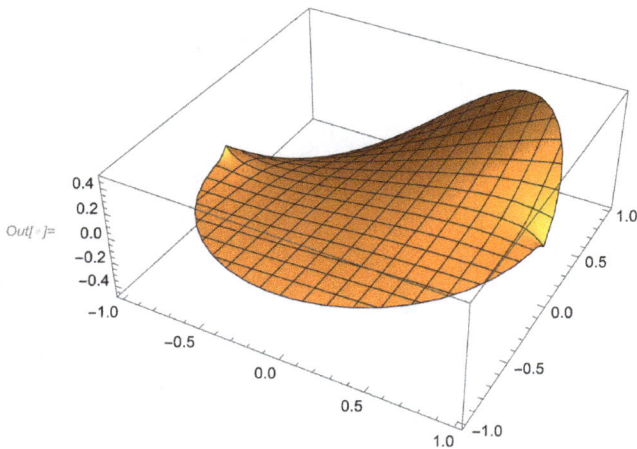

Figure 3.23

```
In[·]:= NDSolve[{Laplacian[u[x, y], {x, y}] == 0,
        DirichletCondition[u[x, y] == 0, x < 0],
        DirichletCondition[u[x, y] == 3, x > 1]},
        u, Element[{x, y}, Disk[]]]
```

Out[·]= {{u → InterpolatingFunction[⊞ 〜 Domain: {{-1., 1.}, {-1., 1.}}
                                           Output: scalar]}}

Figure 3.24

```
In[·]:= Plot3D[Evaluate[u[x, y] /. %],
        Element[{x, y}, Disk[]]]
```

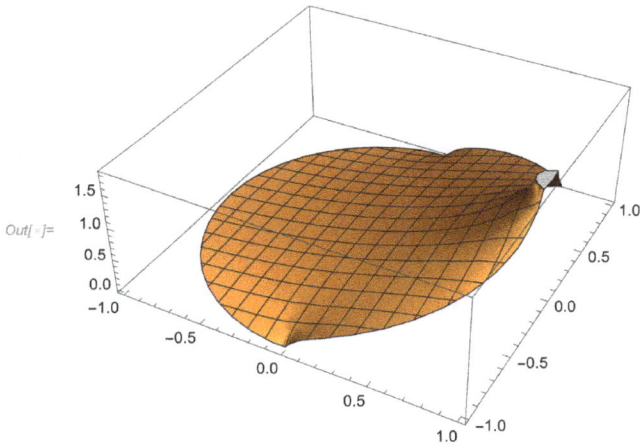

Figure 3.25

4 Differential geometry of curves and surfaces in Mathematica®

There is an excellent book [1] about differential geometry of curves and surfaces which includes Mathematica® code. It contains a detailed account of the theory – definitions, statements and their proofs, and a lot of worked-out examples. Although the code was written for an older version of Mathematica®, it can still be used with some minor modifications even today. In this chapter we rewrote some of the code in a more modern way and created a number of new interactive illustrations. We also discuss the recent changes in Mathematica® relevant to differential geometry of curves and surfaces. We omit most of the theory and formulas since they can be found in [1] or any other textbook on differential geometry, or even Wikipedia. Mathematica® also provides an increasing amount of built-in knowledge as we illustrate in the first section. A lot of information about famous curves and surfaces can be found in [21, 22]. We shall mainly concentrate on aspects of computations and visualization in Mathematica®.

4.1 The Wolfram Knowledgebase

Before we begin with presenting curves and surfaces, we would like to remark that Mathematica® 12 has a lot of built-in famous curves and surfaces, and one can obtain basic information about them. The Wolfram Knowledgebase includes entities of type PlaneCurve, SpaceCurve, Surface, and functions like SpaceCurveData, PlaneCurveData, and SurfaceData. One can retrieve a dataset of all available properties for an entity as seen from several illustrative examples below. We refer the interested reader to the Wolfram Documentation for more information.

The following function gives a big list of built-in plane curves. We omit the output.

In[·]:= Entity["PlaneCurve"] // EntityList;

In[·]:= Length[%]
Out[·]:= 140

A shorter list can be extracted when we specify some properties of plane curves, for instance,

In[·]:= **EntityClass**["PlaneCurve", "AssociatedPeople" → [**Isaac Newton** PERSON]] // **EntityList**

Figure 4.1

https://doi.org/10.1515/9783110774641-004

Out[ ]= { Cartesian ovals , cycloid , cissoid of Diocles , epicycloid ,

epitrochoid , hypocycloid , hypotrochoid , parabola , serpentine curve }

Figure 4.2

The data which is built-in for a given plane curve can be extracted by

In[ ]:= Cartesian ovals PLANE CURVE ["Dataset"]

Figure 4.3

acnodes	–
algebraic degree	4
algebraic equation	Function[{a, k, m, n}, Function[{x, y}, -4 k² n² ((a + x)² + y²) + (k² + 2 a (m² + n²) …
alternate names	{Cartesian curve, ovals of Descartes}
arc length	–
arc length function	–
area enclosed	–
area enclosed plot	0.62, k = 2, m = 7, n 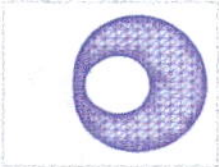
area moment of inertia tensor of enclosed lamina	–
associated entities	{ René Descartes , Isaac Newton }
associated people	{ René Descartes , Isaac Newton }
asymptotes	–
Cartesian equation	Function[{a, k, m, n}, Function[{x, y}, -4 k² n² ((a + x)² + y²) + (k² + 2 a (m² + n²) …
centroid of enclosed lamina	–
Cesàro equation	–
chord length	–
circumference	–
classes	{ ⊞ algebraic curves , ⊞ closed curves , ⊞ ovals , ⊞ polar curves , ⊞ quartic curves }
crunodes	–
parametric curvature	–

⊼ ⋀ rows 1–20 of 80 ⋁ ⋎

Figure 4.4

Some specific information can be obtained as follows:

In[ ]:= ellipse PLANE CURVE ["Area"]

Figure 4.5

Out[ ]= $\text{Function}\left[\{a, b\}, \pi\, a\, b\right]$

Figure 4.6

or, equivalently,

In[ ]:= Entity["PlaneCurve", "Ellipse"]["Area"]

Out[ ]= $\text{Function}\left[\{a, b\}, \pi\, a\, b\right]$

Figure 4.7

In[ ]:= Entity["PlaneCurve", "Ellipse"]["Area"][x, y]
Out[ ]:= Pi*x*y

Sometimes the dataset of an object may not contain required information. For instance, the dataset for the bullet nose curve does not contain its parametric equations, which are $(x, y) = (a \cos t, b \cot t)$:

In[ ]:= Entity["PlaneCurve", "BulletNoseCurve"]["Dataset"]

Out[ ]=

acnodes	—
algebraic degree	4
algebraic equation	$\text{Function}\left[\{a, b\}, \text{Function}\left[\{x, y\}, -b^2 x^2 + a^2 y^2 - x^2 y^2\right]\right]$
alternate names	—
arc length	—
arc length function	—
area enclosed	—
area enclosed plot	—
area moment of inertia tensor of enclosed lamina	—
associated entities	—
associated people	{}
asymptotes	—
Cartesian equation	$\text{Function}\left[\{a, b\}, \text{Function}\left[\{x, y\}, -b^2 x^2 + a^2 y^2 = x^2 y^2\right]\right]$
centroid of enclosed lamina	—
Cesàro equation	—
chord length	—
circumference	—
classes	{ ⊞ algebraic curves , ⊞ quartic curves }
crunodes	Function[a, {{0, 0}}]
parametric curvature	—
⊼ ∧ rows 1–20 of 80 ∨ ⌄	

Figure 4.8

In[·]:= Entity["PlaneCurve", "BulletNoseCurve"]
        ["ParametricEquations"]

Out[·]:= Missing["NotAvailable"]

Other geometric entities available are SpaceCurve and Surface.

In[·]:= Entity["SpaceCurve"] // EntityList

Out[·]= { conchospiral , conical spiral , helix , Seiffert spherical spiral ,

    Pappus spiral , spherical spiral , Steinmetz curve , Viviani curve }

Figure 4.9

To get a bitmap image, we either use

In[·]:= SpaceCurveData["VivianiCurve", "Image"]

Out[·]=

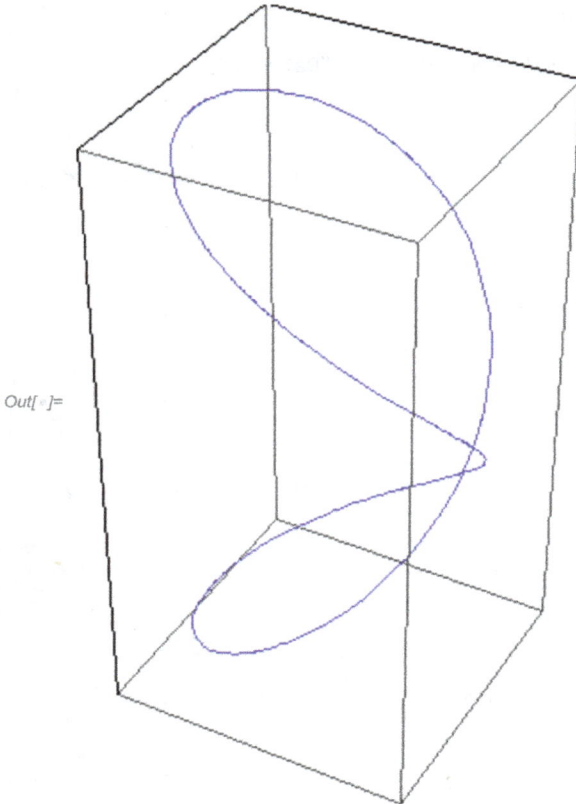

Figure 4.10

or

In[·]:= Entity["SpaceCurve", "VivianiCurve"]["Image"]

In order to visualize the curve (and combine it later with other 3D graphics objects), we can type in

In[·]:= grviv = Entity["SpaceCurve", "VivianiCurve"]
       ["Graphics3D"]

Out[·]=

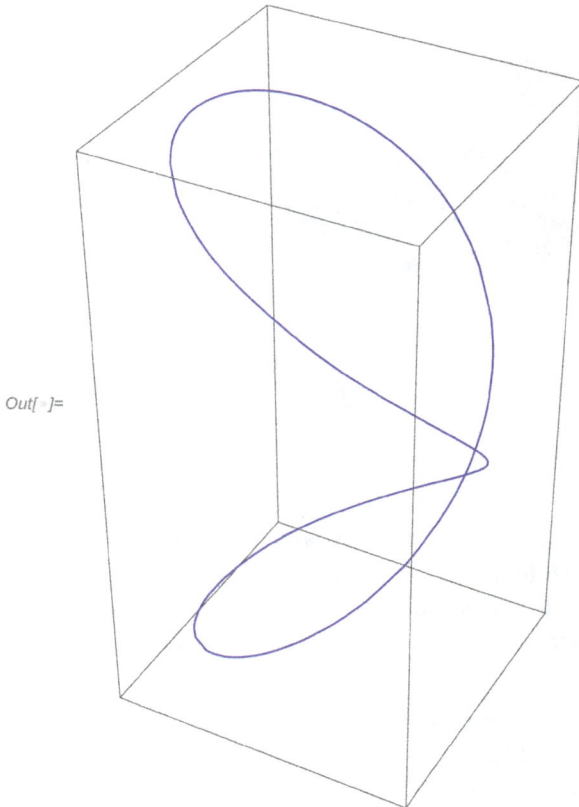

Figure 4.11

In[·]:= Show[{Graphics3D[Ball[{0, 0, 0}, 2]], grviv}]

Out[ ]=

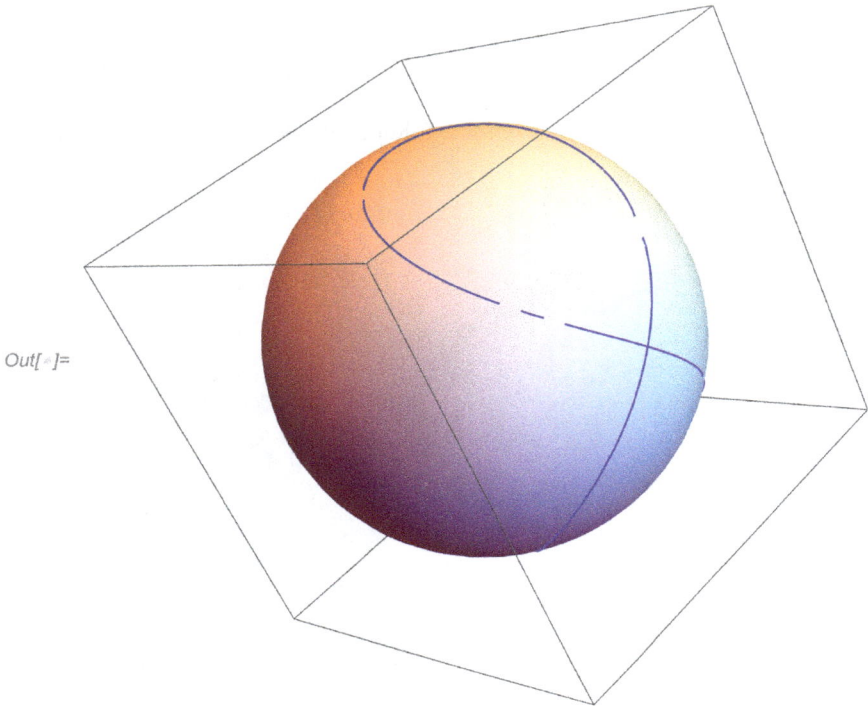

Figure 4.12

In fact, Viviani curve is formed by intersecting a sphere with a cylinder.

On the other hand, if we wish to make some further computations rather than create graphics, then it is preferable to create a Region object. To do this, we need to extract either parametric or implicit equations. In this case we shall use parametric ones.

```
In[·]:= Entity["SpaceCurve", "VivianiCurve"]
         ["ParametricEquations"][a][t]
Out[·]:= {a*(1 + Cos[t]), a*Sin[t], 2*a*Sin[t/2]}
```

We can use parametric equations for further visualization

```
In[·]:= Manipulate[ParametricPlot3D[{a*(1 +
         Cos[t]), a*Sin[t], 2*a*Sin[t/2]},
         {t, 0, 2*Pi}, PlotRange -> {{0, 10},
         {0, 10}, {0, 10}}], {a, 2, 10}]
```

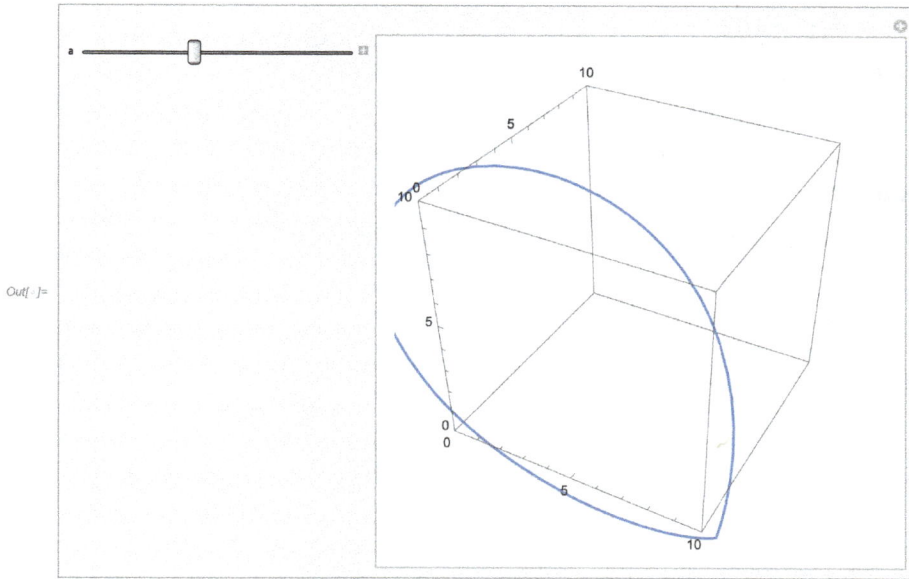

Figure 4.13

or we can create a surface using two parameters:

```
In[·]:= Region[gr1viv = ParametricRegion[{a*(1 +
        Cos[t]), a*Sin[t], 2*a*Sin[t/2]},
        {{a, 0, 1}, {t, 0, 2*Pi}}]]
```

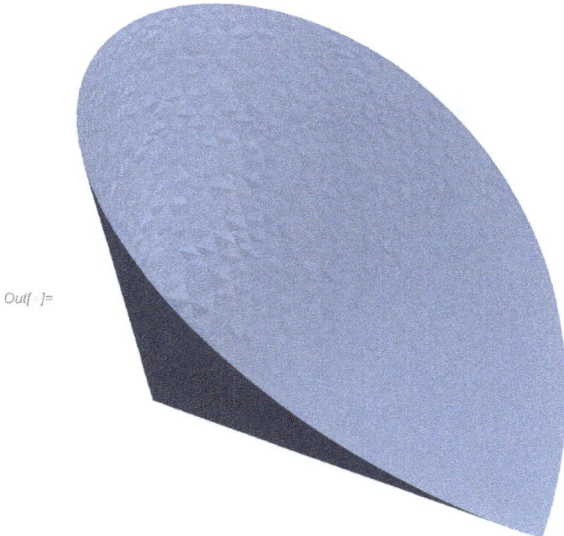

Figure 4.14

and find the surface area:

In[·]:= RegionMeasure[gr1viv]
Out[·]:= 4*Sqrt[2]*EllipticE[1/2]

In[·]:= % // N
Out[·]:= 7.6404

We can extract information about curves in different ways:

In[·]:= **SpaceCurveData**$\left[\boxed{\textbf{helix}\ \text{SPACE CURVE}}, \text{\{"ArcLengthFunction", "Torsion"\}}\right]$

Figure 4.15

Out[·]:= $\left\{\text{Function}\left[\{r, c\}, \text{Function}\left[t, \sqrt{c^2 + r^2}\ t\right]\right], \text{Function}\left[\{r, c\}, \text{Function}\left[t, \dfrac{c}{c^2 + r^2}\right]\right]\right\}$

Figure 4.16

Entities can be used as functions directly rather than arguments of SpaceCurveData. The following inputs return the same result as above:

In[·]:= $\boxed{\textbf{helix}\ \text{SPACE CURVE}}$ **[{"ArcLengthFunction", "Torsion"}]**

Figure 4.17

In[·]:= Entity["SpaceCurve", "Helix"]
        [{"ArcLengthFunction", "Torsion"}];

We again omit the next output which is a long list of surface names:

In[·]:= Entity["Surface"] // EntityList;

In[·]:= Length[%]
Out[·]:= 179

In[·]:= EntityList[EntityClass["Surface",
        "Classes" -> {"Closed"}]]

Out[ ]= {astroidal ellipsoid, Bohemian dome, capsule, citrus surface, closed cone, corner cushion surface, cross-cap, cube surface, cuboid surface, closed cylinder, double sphere, eight surface, ellipsoid, elliptic torus, football surface, Taubin's heart surface, horn torus, Hunt surface, hyperbolic octahedron, lemon surface, Leopold surface, mushroom surface, Nepali surface, oblate spheroid, oloid, piriform surface, prolate spheroid, pumpkin surface, Roman surface, rounded cube surface, sine surface, sphere, spheroid, square pyramid surface, star surface, thistle surface, tooth surface, torus, twilight surface}

Figure 4.18

In[ ]:= spheroid SURFACE ["Dataset"]

Figure 4.19

Out[ ]=

algebraic degree	2
algebraic equation	Function[{a, c}, Function[{x, y, z}, $-a^2 c^2 + c^2 x^2 + c^2 y^2 + a^2 z^2$]]
alternate names	—
area element	Function[{a, c}, Function[{u, v}, $\frac{a \text{Sin}[v] \sqrt{a^2+c^2+(a-c)(a+c)\text{Cos}[2v]}}{\sqrt{2}}$]]
associated people	—
Cartesian equation	Function[{a, c}, Function[{x, y, z}, $\frac{x^2+y^2}{a^2} + \frac{z^2}{c^2} == 1$]]
centroid of solid	Function[{a, c}, {0, 0, 0}]
Christoffel symbol of the second kind	Function[{a, c}, Function[{u, v}, {{{0, Cot[v]}, {Cot[v], 0}}, {{$-\frac{2 a^2 \text{Cos}[v] \text{Sin}[v]}{a^2+c^2+(a^2-c^2)\text{Cos}[2v]}$,
(map) chromatic number	—
classes	{algebraic surfaces, closed surfaces, quadratic surfaces, surfaces of revolution}
cross sections	{circle, ellipse}
entity classes	{algebraic surfaces, closed surfaces, quadratic surfaces, surfaces of revolution}
Euler characteristic	2
filled region	—
coefficients of the first fundamental form	Function[{a, c}, Function[{u, v}, {$a^2 \text{Sin}[v]^2$, 0, $\frac{1}{2}(a^2 + c^2 + (a-c)(a+c)\text{Cos}[2v])$}]]
Gaussian curvature	Function[{a, c}, Function[{u, v}, $\frac{4 c^2}{(a^2+c^2+(a-c)(a+c)\text{Cos}[2v])^2}$]]
generalized diameter	—
genus	0
3-D graphics	...
image	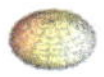

rows 1–20 of 50

Figure 4.20

We can get the list of properties by typing

In[·]:= `Entity["Surface", "Ellipsoid"]["Properties"]`

Out[·]= { algebraic degree , algebraic equation , alternate names , area element , associated people , Cartesian equation , centroid of solid , Christoffel symbol of the second kind , (map) chromatic number , classes , cross sections , entity classes , Euler characteristic , filled region , coefficients of the first fundamental form , Gaussian curvature , generalized diameter , genus , 3-D graphics , image , implicit Gaussian curvature , implicit mean curvature , implicit normal vector , moment of inertia tensor of solid , CDF of lengths , mean line segment length , PDF of lengths , mean curvature , metric tensor , name , number of nodes , normal vector , parameter range , parameters , parametric equations , principal curvatures , number of punctures , related Wolfram Language symbols , Ricci tensor , Riemann tensor , coefficients of the second fundamental form , semialgebraic description , singular points , sport objects , surface area , variable constraints , variable descriptions , variables , vector length , volume of solid }

Figure 4.21

Some information for surfaces is available, and some is not:

In[·]:= Entity["Surface", "Spheroid"]["AlternateNames"]
Out[·]:= Missing["NotAvailable"]

In[·]:= Entity["Surface", "Ellipsoid"]["AlgebraicDegree"]
Out[·]:= 2

In[·]:= Entity["Surface", "Ellipsoid"]["GaussianCurvature"]

Out[·]= $\text{Function}\left[\{a, b, c\}, \text{Function}\left[\{u, v\}, \dfrac{a^2\, b^2\, c^2}{(a^2\, b^2\, \text{Cos}[v]^2 + c^2\, (b^2\, \text{Cos}[u]^2 + a^2\, \text{Sin}[u]^2)\, \text{Sin}[v]^2)^2}\right]\right]$

Figure 4.22

In[·]:= Entity["Surface", "Ellipsoid"]
        ["GaussianCurvature"][1, 1, 1][x, y]
Out[·]:= 1/(Cos[y]^2 + (Cos[x]^2 + Sin[x]^2)*Sin[y]^2)^2

In[·]:= SurfaceData["SampleEntities"]

Out[·]= { astroidal ellipsoid , infinite cone , crixxi surface , finite cylinder , finite horizontal cylindrical segment , finite elliptic cylinder , plane , seashell surface , sphere , spheroid }

Figure 4.23

In[·]:= SurfaceData["KissSurface", "Image"]

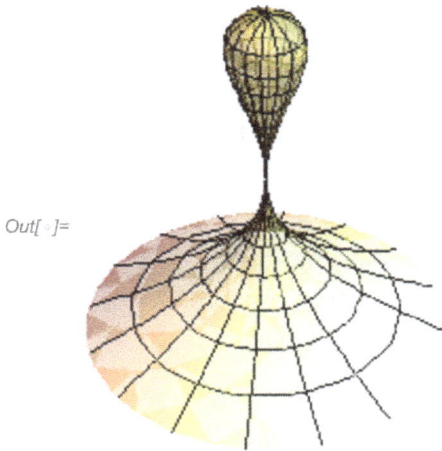

Out[ ]=

Figure 4.24

In[ ]:= **SurfaceData** $\Big[$ spheroid SURFACE $\Big]$, {"Volume", "InertiaTensor"} $\Big]$

Figure 4.25

Out[ ]= $\Big\{ \text{Function}\Big[\{a, c\}, \frac{4}{3}\pi a^2 c\Big], \text{Function}\Big[\{a, c\}, \frac{1}{5}(a^2 + c^2) \text{IdentityMatrix}[3]\Big]\Big\}$

Figure 4.26

4.2 Vectors in Mathematica®

Before we start our presentation of curves and surfaces, let us revisit the notions of vectors, their norms, and other operators on vectors in Mathematica®. We have already explained most of the following material in Volume 2 [8].

A vector in Mathematica® is represented by a list of numbers or by a list of variables which are assumed to take numerical values. We can check whether an object is a vector or not by using the function VectorQ. For example,

In[·]:= Map[VectorQ, {{a, 1, I}, {a, 2, {2, 3}}}]
Out[·]:= {True, False}

In[·]:= VectorQ /@ {{a, 1, I}, {a, 2, {2, 3}}}
Out[·]:= {True, False}

The length of a vector is called its norm. **Mathematica®** has a built-in function Norm. In symbolic expressions it assumes that all the variables are complex:

In[·]:= Norm[{a, b, 1, I}]
Out[·]:= Sqrt[2 + Abs[a]^2 + Abs[b]^2]

The built-in function Norm uses absolute values in the answer since it treats *a* and *b* as complex and one needs additional commands to remove them. It is possible to deal with this by using Simplify with Assuming as follows:

In[·]:= Assuming[Element[{a, b}, Reals],
        Simplify[Norm[{a, b, 1, I}]]]
Out[·]:= Sqrt[2 + a^2 + b^2]

But it is quicker to use the function ComplexExpand:

In[·]:= ComplexExpand[Norm[{a, b, 1, I}]]
Out[·]:= Sqrt[2 + a^2 + b^2]

We could also define our own function norm (we distinguish the name using the small letters):

In[·]:= norm[(v_)?VectorQ] := Sqrt[v . v]

In[·]:= norm[{a, b, c}]
Out[·]:= Sqrt[a^2 + b^2 + c^2]

One more useful built-in function is the function Normalize:

In[·]:= Normalize[{3, 0, 4}]
Out[·]:= {3/5, 0, 4/5}

The operations on vectors that we shall need in the course of differential geometry are the dot and cross products. The dot product

In[·]:= Dot[{a, b, c}, {e, f, g}]
Out[·]:= a e + b f + c g

gives a scalar and can also be used with a dot (.) as follows:

In[·]:= {1, -1} . {1, 1}
Out[·]:= 0

We also multiply matrices using Dot:

In[·]:= {{1, 1}, {1, 1}} . {{1, 0}, {0, 1}}
Out[·]:= {{1, 1}, {1, 1}}

The vector product, on the other hand, gives a vector:

In[·]:= Cross[{a, b, c}, {d, e, f}]
Out[·]:= {-c e + b f, c d - a f, -b d + a e}

It is defined for $\mathbb{R}^3$ vectors. The × sign can be easily entered by typing Esc + cross + Esc.

As we already know from Volume 2, Mathematica® can verify the triangle and Cauchy–Schwarz inequalities, but only for a small number of variables:

In[·]:= `Reduce[ForAll[{u, v, p, q}, Element[{`
        `u, v, p, q}, Reals], Norm[{u, v} + {p, q}] <=`
        `Norm[{u, v}] + Norm[{p, q}]]]`
Out[·]:= True

In[·]:= `Reduce[ForAll[{u, v, p, q}, Element[{`
        `u, v, p, q}, Reals], Abs[{u, v} .`
        `{p, q}] <= Norm[{p, q}]*Norm[{u, v}]]]`
Out[·]:= True

Instead of Reduce we could use FullSimplify:

In[·]:= `FullSimplify[ForAll[{u, v, p, q}, Element[{`
        `u, v, p, q}, Reals], Norm[{u, v} + {p, q}]`
        `<= Norm[{u, v}] + Norm[{p, q}]]]`
Out[·]:= True

In[·]:= `FullSimplify[ForAll[{u, v, p, q}, Element[{`
        `u, v, p, q}, Reals], Abs[{u, v} . {p, q}]`
        `<= Norm[{p, q}]*Norm[{u, v}]]]`
Out[·]:= True

Mathematica® can also work with abstract (symbolic) tensors:

In[·]:= `TensorExpand[v . (w + u)]`
Out[·]:= `v . u + v . w`

In[·]:= `TensorExpand[Cross[v, w + u]]`
Out[·]:= `-Cross[u, v] + Cross[v, w]`

In[·]:= `TensorExpand[v . Cross[v, w]]`
Out[·]:= `0`

4.3 Curves

4.3.1 A physical interpretation of a curve in $\mathbb{R}^n$

A continuous mapping $f : I \to \mathbb{R}^n$, where $n \geq 2$ and I is an interval, is called a curve. When $n = 2$, the curve is called a plane (planar) curve. When $n = 3$, it is a space (spacial) curve. These can be represented graphically by means of Mathematica®'s functions ParametricPlot and ParametricPlot3D. For example, the function $f : [0, 2\pi] \to \mathbb{R}^2$ given by $f(t) = (x(t), y(t)) = (t \cos t, t \sin t)$ represents a spiral:

In[·]:= `ParametricPlot[{t Cos[t], t Sin[t]}, {t, 0, 2 Pi}]`

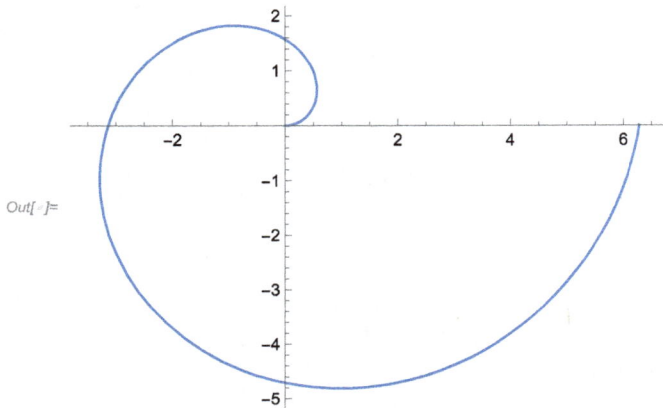

Figure 4.27

It is convenient to think of *t* as time. The function gives the equation of motion of a particle in a given time interval.

```
In[·]:= Manipulate[Module[{spiral = ParametricPlot[{
        s*Cos[s], s*Sin[s]}, {s, 0, 2*Pi}]}, Show[{
        Graphics[{Red, PointSize[0.02], Point[{
        t*Cos[t], t*Sin[t]}]}], spiral}]],
        {t, 0, 2*Pi, Appearance -> "Labeled"}]
```

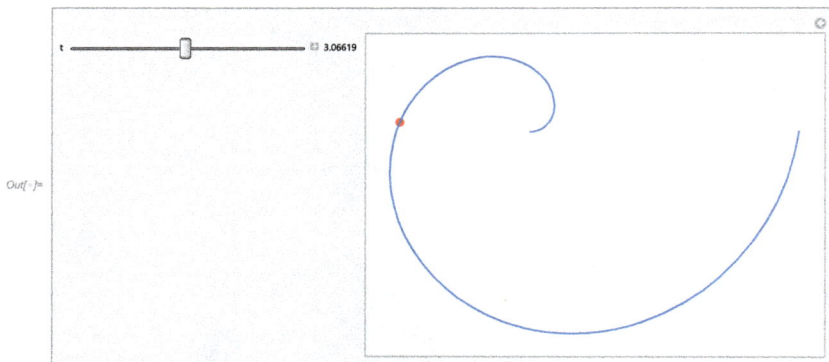

Figure 4.28

We can use a different approach to this when we intent to perform some automatic computations. We can start by defining a parametric region

```
In[·]:= spiral1 = ParametricRegion[{t Cos[t],
        t Sin[t]}, {{t, 0, 2 Pi}}]
Out[·]= ParametricRegion[{{Cos[t]*t, t*Sin[t]},
        0 <= t <= 2*Pi}, {t}]
```

We can now visualize this by

In[·]:= Region[spiral1, Axes -> True]

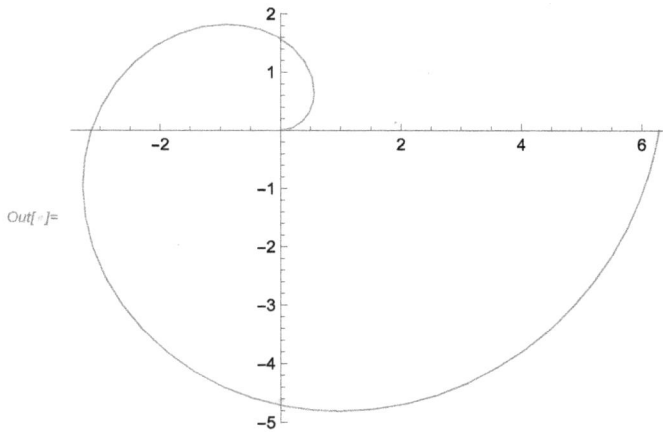

Out[·]=

Figure 4.29

We can compute the length of this spiral with RegionMeasure:

In[·]:= RegionMeasure[spiral1]
Out[·]:= Pi*Sqrt[1 + 4*Pi^2] + (1/2)*ArcSinh[2*Pi]

Unlike with graphics we cannot show one region on top of the other, but we can join them. For example,

In[·]:= Region[RegionUnion[spiral1, Point[{0.7, 0.2}]]]

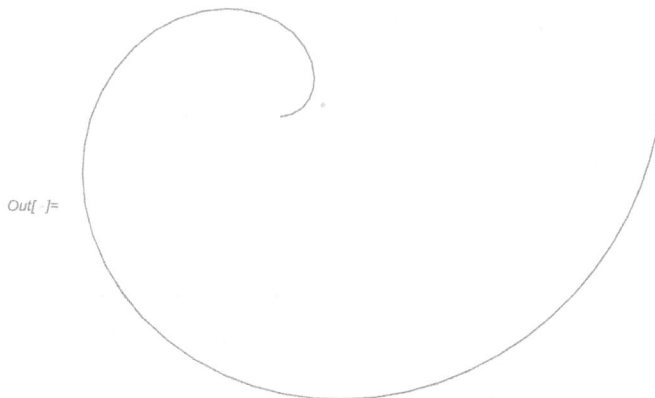

Out[·]=

Figure 4.30

We can change the size of a point, but we cannot change the color of it independently:

```
In[·]:= spp[s_] := RegionUnion[ParametricRegion[{
        t*Cos[t], t*Sin[t]}, {{t, 0, 2*Pi}}],
        Point[{s*Cos[s], s*Sin[s]}]]]
```

```
In[·]:= Manipulate[Region[Style[spp[t],
        PointSize[0.02], Red]], {t, 0, 2*Pi}]
```

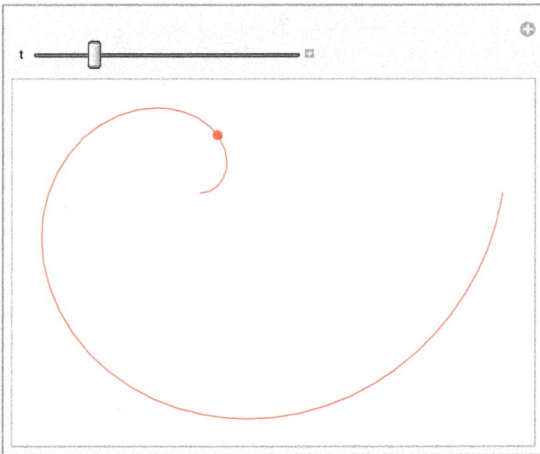

Figure 4.31

The advantage of the approach with Region is that we can further perform various computations with an object, for instance, find the distance from a given point:

```
In[·]:= RegionDistance[spiral1, {10, 10}]//N
Out[·]:= 10.6684
```

or the nearest point:

```
In[·]:= RegionNearest[spiral1, {10, 10}]
Out[·]:= {2*Pi, 0}
```

Now suppose we have a function $f : [0, 2\pi n] \rightarrow \mathbb{R}^3, t \mapsto (R\cos t, R\sin t, at/(2\pi))$. In the code below we shall use the name spiralpar instead of f. We think of it as a function of one variable (e. g., time) with 3 parameters n, R, a and want to find a geometric interpretation for the parameters. Of course, we can view f as a function of four variables; the distinction between "variables" and "parameters" is a matter of interpretation. Hence we define

```
In[·]:= spiralpar[n_, R_, a_][t_] := {R*Cos[t],
        R*Sin[t], a*(t/(2*Pi))}
```

In[·]:= Manipulate[Show[ParametricPlot3D[
      spiralpar[n, R, a][t], {t, 0, 2*n*Pi}],
      Graphics3D[{Blue, Arrow[Tube[{spiralpar[
      n, R, a][s], spiralpar[n, R, a][s] +
      Derivative[1][spiralpar[n, R, a]][s]}]],
      Red, Point[spiralpar[n, R, a][s]]}], PlotRange
      -> {{-12, 12}, {-12, 12}, {0, 18}}],
      {{s, Pi/2, "t"}, 0, 2*n*Pi, Appearance ->
      "Labeled"}, {{R, 2, "R"}, 0, 4, Appearance
      -> "Labeled"}, {{a, 1, "a"}, 0.1, 2, Appearance
      -> "Labeled"}, {{n, 1, "n"}, 1, 5, 1,
      Appearance -> "Labeled"}]

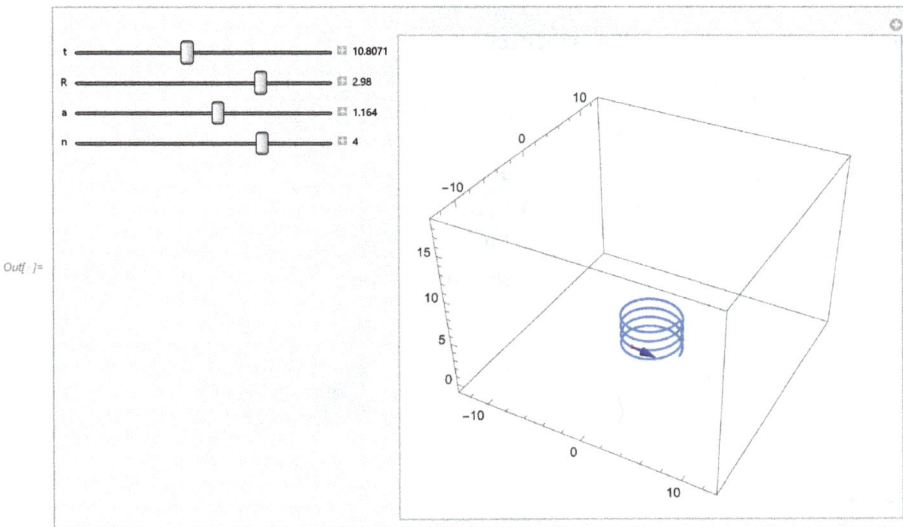

Figure 4.32

We can interpret the function f (spiralpar) as giving the position of a particle moving in such a way that its path forms a spiral. The parameter R is the radius of the circle on which the projection of the particle onto the x–y plane moves, the parameter a is the vertical distance which the point travels while making one complete rotation. The arrow represents the velocity of the moving point. Mathematically it is a vector-valued function which is the derivative of the vector-valued position function f.

    Let us have a look at another example.

In[·]:= Entity["PlaneCurve", "AgnesiWitch"]
      ["ParametricEquations"][2][t]
Out[·]:= {4*t, 4/(1 + t^2)}

We can also define the witch of Agnesi curve by

In[·]:= `agnesi[a_][t_] := {2*a*Tan[t], 2*a*Cos[t]^2};`

In[·]:= `ParametricPlot[{{4*t, 4/(1 + t^2)},`
`        agnesi[2][t]}, {t, -1, 1}]`

Out[·]=

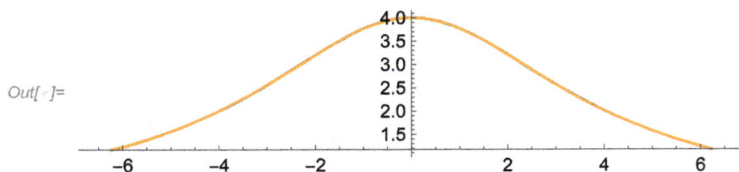

Figure 4.33

In[·]:= `Manipulate[Show[{ParametricPlot[`
`        agnesi[1][t], {t, -Pi, Pi}], Graphics[{Red,`
`        PointSize[0.02], Point[agnesi[1][t]],`
`        Arrow[{agnesi[1][t], agnesi[1][t] +`
`        D[agnesi[1][s], s] /. s -> t}]}], Axes ->`
`        True, PlotRange -> {{-20, 20},`
`        {-2, 2}}]}], ImageSize -> Large],`
`        {{t, 0, "t"}, -Pi, Pi, Appearance`
`        -> "Labeled"}]`

Out[·]=

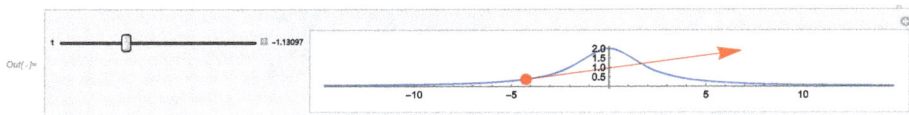

Figure 4.34

We now see that the particle following the parametric equation of motion first moves to the right with increasing speed and then at $t = \pi/2$ reaches ∞ and then returns back from $-\infty$. Normalizing the tangent vector, we have the following illustration:

In[·]:= `Manipulate[Show[{ParametricPlot[`
`        agnesi[1][t], {t, -Pi, Pi}], Graphics[{Red,`
`        PointSize[0.01], Point[agnesi[1][t]],`
`        Arrow[{agnesi[1][t], agnesi[1][t] +`
`        Normalize[D[agnesi[1][s], s] /. s -> t]}]}],`
`        Axes -> True, PlotRange -> {{-10, 10},`
`        {-2, 2}}]}], {t, 0, Pi, Appearance ->`
`        "Labeled"}]`

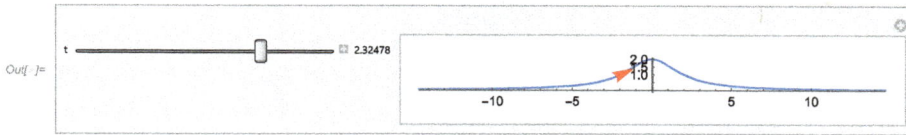

Figure 4.35

Note that in the code above we can use expression `Normalize[agnesi[1]'[t]]` instead of `Normalize[D[agnesi[1][s],s]/.s->t]`.

4.3.2 Implicitly defined curves

The curves in $\mathbb{R}^2$ may be given not only by parametric equations. They can also be given implicitly by an equation $F(x, y) = 0$. In this case we can visualize them using `ContourPlot`:

```
In[·]:= Entity["PlaneCurve", "CassiniOvals"]
        ["AlgebraicEquation"][a, b][x, y]
Out[·]:= a^4 - b^4 - 2*a^2*x^2 + x^4 +
        2*a^2*y^2 + 2*x^2*y^2 + y^4
```

```
In[·]:= cassini[a_, b_][x_, y_] := (x^2 +
        y^2 + a^2)^2 - b^4 - 4*a^2*x^2
```

```
In[·]:= ContourPlot[cassini[2.01, 2][x, y] == 0,
        {x, -4, 4}, {y, -2, 2}]
```

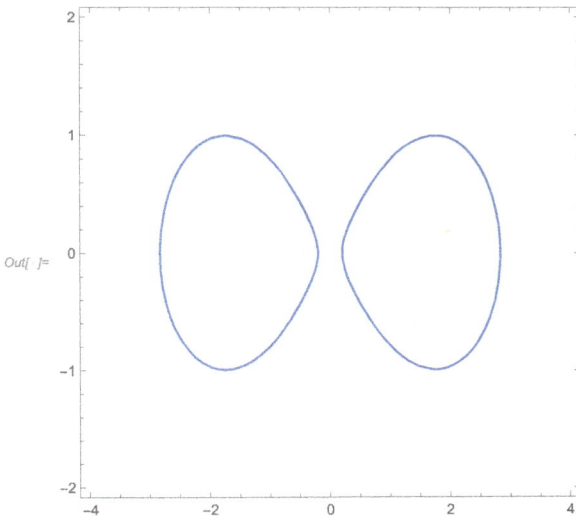

Figure 4.36

In[·]:= Manipulate[ContourPlot[cassini[a, b][x, y]
        == 0, {x, -4, 4}, {y, -2, 2}], {{a, 2.01, "a"},
        -4, 4, Appearance -> "Labeled"}, {{b, 2., "b"},
        -4, 4, Appearance -> "Labeled"}]

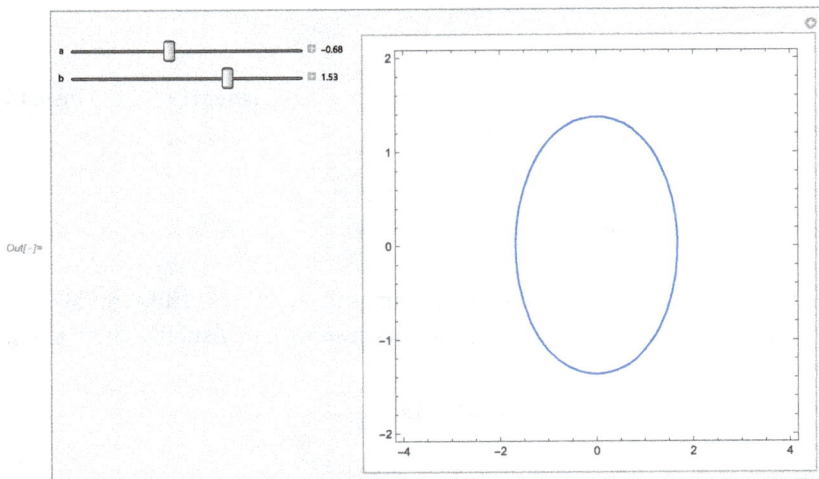

Figure 4.37

Let us consider another example.

In[·]:= ContourPlot[Cos[x] + Cos[y] == 1/2,
        {x, 0, 4*Pi}, {y, 0, 4*Pi}]

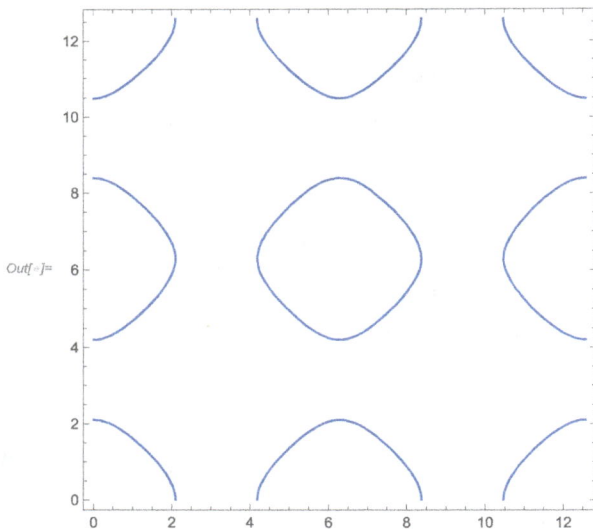

Figure 4.38

Another way is to use ImplicitRegion which allows us to compute the total length of all the pieces and many other things:

In[·]:= Region[ImplicitRegion[Cos[x] + Cos[y] == 1/2,
      {{x, 0, 4 Pi}, {y, 0, 4 Pi}}]]

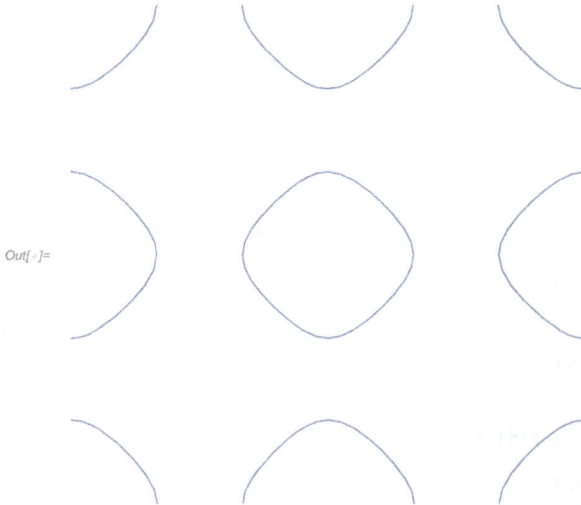

Figure 4.39

In[·]:= RegionMeasure[%]
Out[·]:= 50.1034

However, as shown below, Mathematica® does not always work properly with infinite intervals in this example:

In[·]:= impreg = Region[ImplicitRegion[Cos[x] +
      Cos[y] == 1/2, {{x, -Infinity, Infinity},
      {y, -Infinity, Infinity}}]]

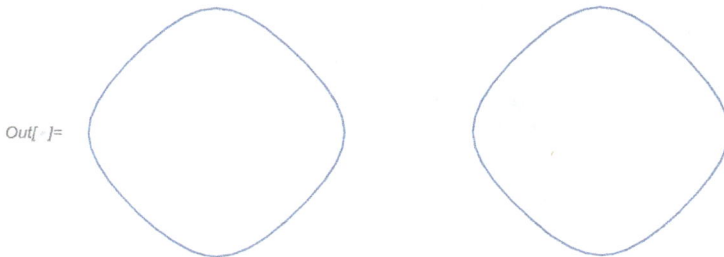

Figure 4.40

but this gives a correct answer:

In[·]:= impreg1 = Reduce[RegionMember[
  ImplicitRegion[Cos[x] + Cos[y] ==
  1/2, {{x, -Infinity, Infinity}, {y, -Infinity,
  Infinity}}], {x, y}], {x, y}, Reals]

Out[·]= $(c_1 \mid c_2) \in \mathbb{Z}$ && $\frac{1}{3} \times (-2\pi + 6\pi c_2) \le x \le \frac{1}{3} \times (2\pi + 6\pi c_2)$ &&

$\left(y == -\text{ArcCos}\left[\frac{1}{2} \times (1 - 2\text{Cos}[x]) \right] + 2\pi c_1 \mid\mid y == \text{ArcCos}\left[\frac{1}{2} \times (1 - 2\text{Cos}[x]) \right] + 2\pi c_1 \right)$

Figure 4.41

In[·]:= RegionMeasure[impreg]
Out[·]:= 12.5258

In[·]:= BoundedRegionQ[impreg]
Out[·]:= True

In[·]:= BoundedRegionQ[impreg1]
Out[·]:= False

In the following example everything is correct:

In[·]:= Region[ImplicitRegion[x*y == 1, {{x,
  -Infinity, Infinity}, {y, -Infinity,
  Infinity}}]]

Out[·]=

Figure 4.42

In[·]:= RegionMeasure[%]
Out[·]:= Infinity

In[·]:= BoundedRegionQ[%%]
Out[·]:= False

4.3.3 Parametric plots of curves in 2D and 3D

As mentioned above, a parametric curve in $\mathbb{R}^n$ is defined as the image of a smooth function $r : I \rightarrow \mathbb{R}^n$, where I is some interval (not necessarily finite). To plot a curve in $\mathbb{R}^3$, we use the built-in **Mathematica**® function ParametricPlot3D. In $\mathbb{R}^2$ we use ParametricPlot.

We can use ParametricPlot for complicated functions, e. g., for piecewise functions:

In[·]:= pw[t_] := Piecewise[{{{2*Cos[t], 2*Sin[t]},
        0 <= t <= 2*Pi}, {{Cos[t], Sin[t]},
        2*Pi <= t <= 4*Pi}}]

In[·]:= ParametricPlot[pw[t], {t, 0, 4*Pi}]

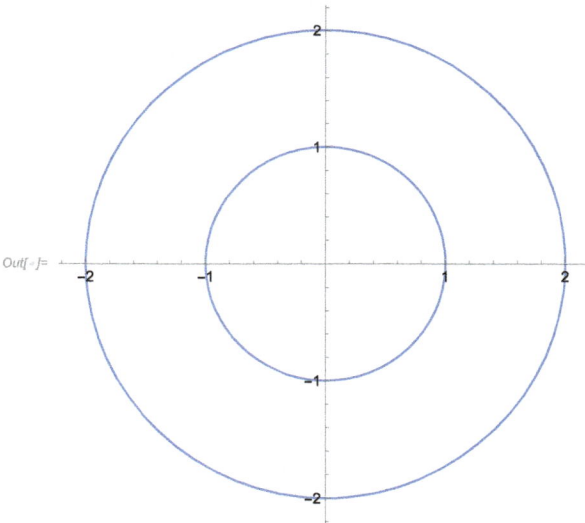

Figure 4.43

The same curve may be defined by more than one parametrization.

In[·]:= r1[t_] := {Cos[t], Sin[t], t}

In[·]:= r2[t_] := {(1 - t^2)/(t^2 + 1),
        (2*t)/(t^2 + 1), 2*ArcTan[t]}

In[·]:= pr1 = ParametricPlot3D[r1[t], {t, 0, Pi},
        PlotStyle -> Directive[Opacity[0.5], Blue]];

In[·]:= `pr2 = ParametricPlot3D[r2[t], {t, 0, Pi},`
    `PlotStyle -> Directive[Opacity[0.5], Green]];`

We can use the function Grid to show the two curves next to each other, but their relationship is not made clear:

In[·]:= `Grid[{{pr1, pr2}}]`

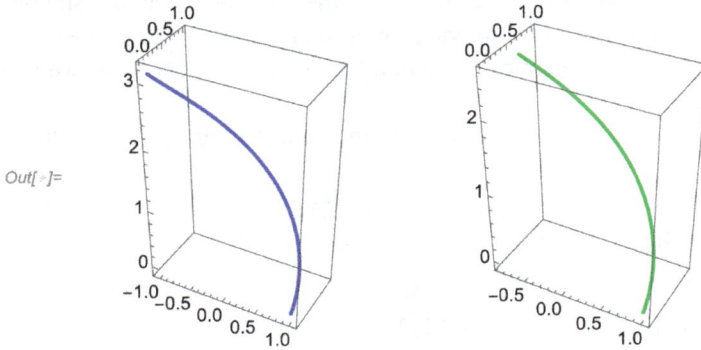

Out[·]=

Figure 4.44

We can use Show to combine them into one picture:

In[·]:= `Show[pr1, pr2]`

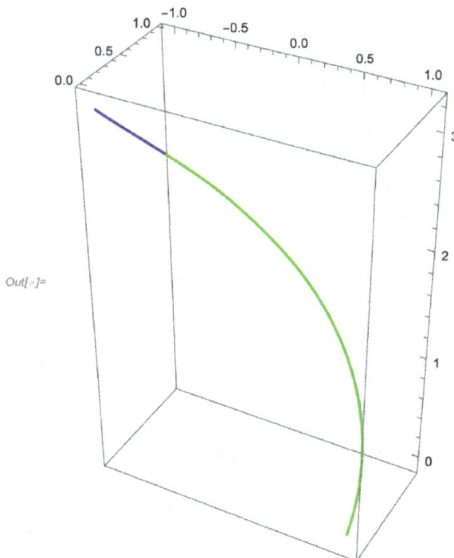

Out[·]=

Figure 4.45

Clearly, the two parametrizations describe the same curve.

Next we shall examine how Mathematica® treats different parametrizations using Regions. There seems to be a bug in Mathematica® that makes it sometimes unable to recognize that two different descriptions of a line give the same region. That bug will hopefully be fixed in the next version of Mathematica®, but the example below gives us a chance to demonstrate a number of useful functions (even though the returned answers are contradictory and sometimes incorrect).

```
In[·]:= rr3 = ParametricRegion[{t, -t},
          {{t, -Infinity, Infinity}}]
Out[·]:= ParametricRegion[{t, -t}, {t}]
```

```
In[·]:= rr4 = ImplicitRegion[x + y == 0, {{x,
          -Infinity, Infinity}, {y, -Infinity, Infinity}}]
Out[·]:= ImplicitRegion[x + y == 0, {x, y}]
```

```
In[·]:= RegionEqual[rr3, rr4]
Out[·]:= False
```

```
In[·]:= RegionEqual[DiscretizeRegion[rr3],
          DiscretizeRegion[rr4]]
Out[·]:= True
```

```
In[·]:= DiscretizeRegion[rr3]
```

Out[·]=

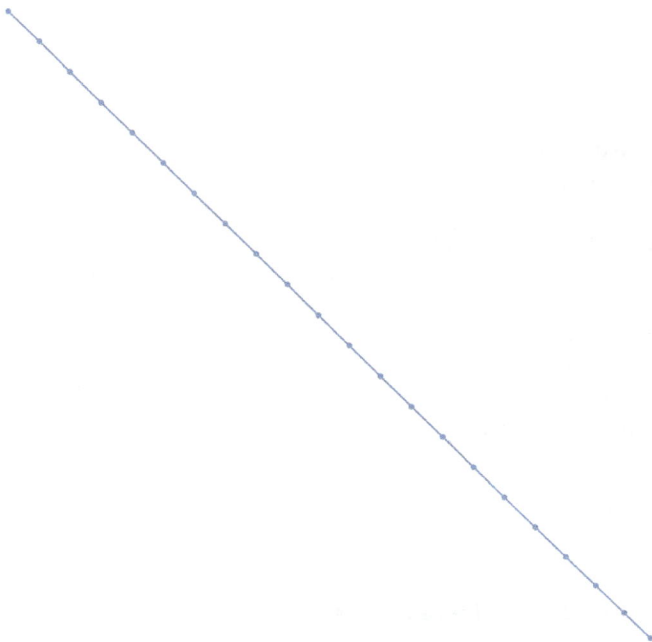

Figure 4.46

In[·]:= `Region[RegionDifference[rr4, rr3],`
        `Axes -> True]`

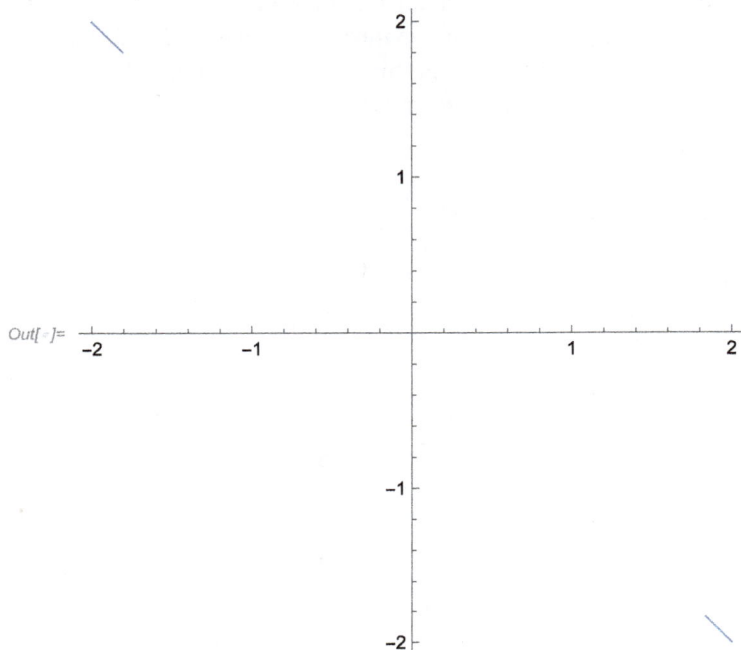

Figure 4.47

In[·]:= `{RegionMeasure[rr3], RegionMeasure[rr4]}`
Out[·]:= `{Infinity, Infinity}`

In[·]:= `RegionMember[rr3, {x, y}]`
Out[·]:= `Element[x |y, Reals] && x + y == 0`

In[·]:= `RegionMember[rr4, {x, y}]`
Out[·]:= `Element[x |y, Reals] && x + y == 0`

In[·]:= `RegionWithin[rr3, rr4]`
Out[·]:= `False`

In[·]:= `RegionWithin[rr4, rr3]`
Out[·]:= `True`

In[·]:= `RegionMember[RegionDifference[rr4,`
        `rr3], {x, y}]`
Out[·]:= `False`

In[·]:= RegionEqual[EmptyRegion[2],
      RegionSymmetricDifference[rr3, rr4]]

Out[·]:= True

Returning to our previous example, Mathematica® cannot decide whether we have the same curve or not:

In[·]:= rr1 = ParametricRegion[r1[t],
      {{t, -Infinity, Infinity}}]

Out[·]:= ParametricRegion[{Cos[t], Sin[t], t}, {t}]

In[·]:= rr2 = ParametricRegion[r2[t],
      {{t, -Infinity, Infinity}}]

Out[·]:= ParametricRegion[{(1 - t^2)/(1 + t^2),
      (2*t)/(1 + t^2), 2*ArcTan[t]}, {t}]

In[·]:= RegionEqual[rr1, rr2]

Out[·]:= RegionEqual$\left[\text{ParametricRegion}[\{\text{Cos}[t], \text{Sin}[t], t\}, \{t\}], \text{ParametricRegion}\left[\left\{\left\{\frac{1-t^2}{1+t^2}, \frac{2t}{1+t^2}, 2\,\text{ArcTan}[t]\right\}, \{t\}\right\}\right]\right]$

Figure 4.48

In[·]:= RegionEqual[DiscretizeRegion[rr1],
      DiscretizeRegion[rr2]]

Out[·]:= False

In[·]:= {Region[rr1, Axes -> True],
      Region[rr2, Axes -> True]}

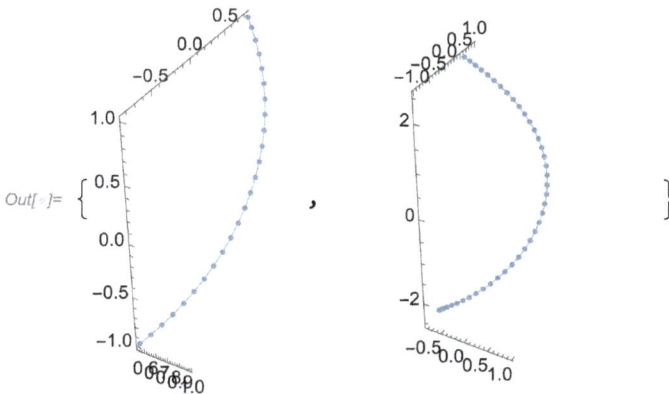

Figure 4.49

However, symbolically, we see that

```
In[·]:= {Cos[t], Sin[t]} /. t -> 2 ArcTan[s]
         // TrigExpand // Simplify
Out[·]:= {(1 - s^2)/(1 + s^2), (2*s)/(1 + s^2)}
```

We can see the dynamics of the motion, which is created using the **Mathematica**® function Manipulate. Note that we used the function If to stop the red point, otherwise it goes out of the picture.

```
In[·]:= Manipulate[Show[Graphics3D[{PointSize[0.02],
         Red, If[t <= Pi, Point[r1[t]], Point[r1[Pi]]],
         Purple, Point[r2[t]]}], pr1], {t, 0, 20,
         Appearance -> "Labeled"}]
```

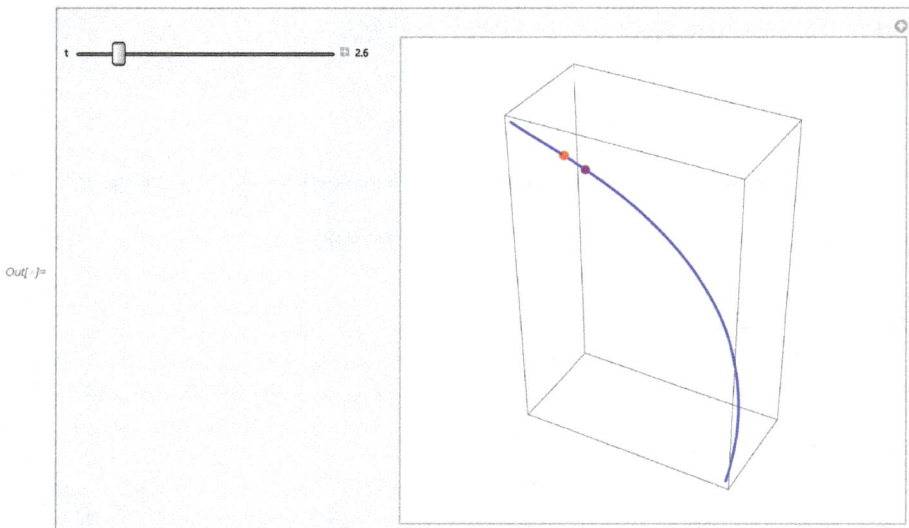

Figure 4.50

The following demonstration shows the same points with velocity vectors:

```
In[·]:= Manipulate[Show[Graphics3D[{
         PointSize[0.02], Red, If[t <= Pi,
         {Point[r1[t]], Arrow[{r1[t], r1[t] +
         Derivative[1][r1][t]}]}, Point[r1[Pi]]],
         Purple, Point[r2[t]], Arrow[{r2[t], r2[t] +
         Derivative[1][r2][t]}]}], pr1, PlotRange
         -> {{-1, 1}, {-1, 1}, {0, 3}}],
         {t, 0, 20, Appearance -> "Labeled"}]
```

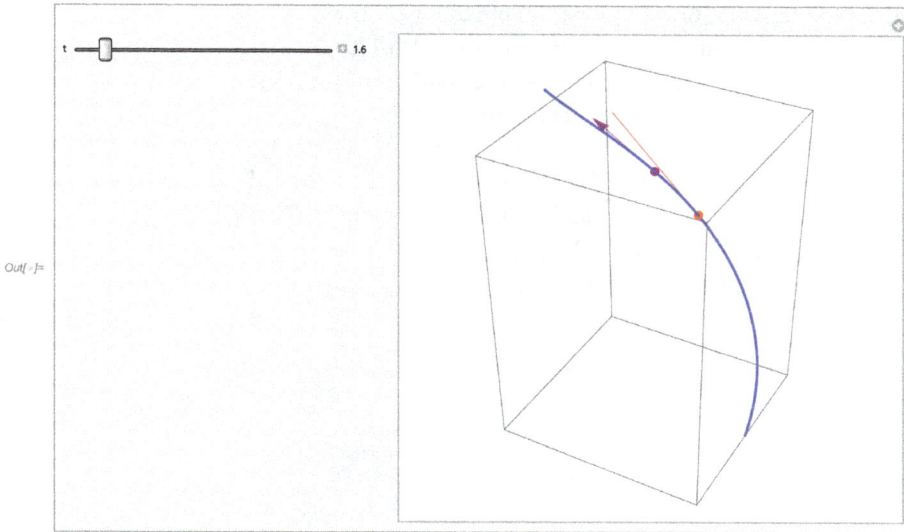

Figure 4.51

4.3.4 Length of parametric curves and the function ArcLength

To calculate the length of the parametric curve on the interval $t \in [a, b]$ (either exactly or numerically), we can define the following functions:

```
In[·]:= curveLength[g_, t_, {a_, b_}] :=
          Integrate[Norm[D[g, t]], {t, a, b}]
```

```
In[·]:= curveLengthN[g_, t_, {a_, b_}] :=
          NIntegrate[Norm[D[g, t]], {t, a, b}]
```

For the curve defined by either of the following two parametrizations

```
In[·]:= r1[t_] := {Cos[t], Sin[t], t}
```

```
In[·]:= r2[t_] := {(1 - t^2)/(t^2 + 1),
          (2*t)/(t^2 + 1), 2*ArcTan[t]}
```

we have

```
In[·]:= curveLengthN[r1[t], t, {0, Pi}]
Out[·]:= 4.44288
```

```
In[·]:= curveLengthN[r2[t], t, {0, Infinity}]
Out[·]:= 4.44288
```

We can see how the length is changing for both parametrizations of the curve defined above:

In[·]:= `Manipulate[Grid[{{Show[Graphics3D[{PointSize[`
`0.02], Red, If[t <= Pi, {Point[r1[t]],`
`Arrow[{r1[t], r1[t] + Derivative[1][r1][t]}]},`
`Point[r1[Pi]]], Purple, Point[r2[t]],`
`Arrow[{r2[t], r2[t] + Derivative[1][r2][t]}]}],`
`pr1], Style[curveLengthN[r1[s], s,`
`{0, Min[Pi, t]}], Red], Style[curveLengthN[r2[s],`
`s, {0, t}], Blue]}}], {t, 0, 20,`
`Appearance -> "Labeled"}]`

Out[·]=

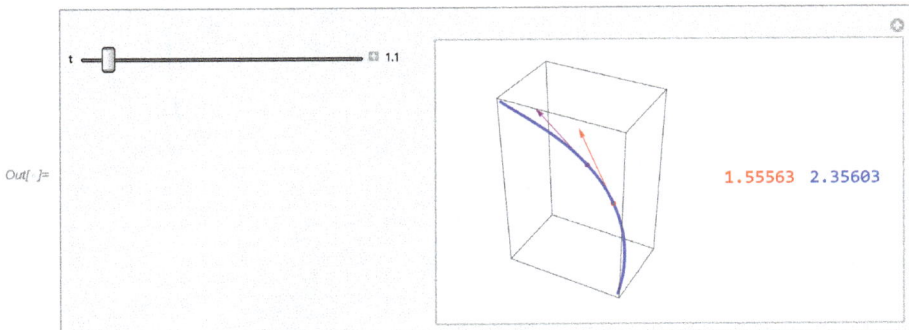

Figure 4.52

Here pr1 was defined in the previous subsection.

There is a built-in function ArcLength that gives the length of the one-dimensional object or the parametrized curve:

In[·]:= `ArcLength[Line[{{0, 0}, {1, 1}, {2, 0}}]]`
Out[·]:= `2*Sqrt[2]`

In[·]:= `ArcLength[Circle[{x, y}, r]]`
Out[·]:= `2*Pi*r`

In[·]:= `ArcLength[{Cos[t], Sin[t], t}, {t, 0, Pi}]`
Out[·]:= `Sqrt[2]*Pi`

What is usually called the length of a curve is a special case of the more general concept of measure, so we can equally well use the more general function RegionMeasure (which also calculates areas, volumes, etc.). We may specify that we are computing a one-dimensional measure but **Mathematica**® guesses correctly the dimension of the object to which we are applying it.

In[·]:= `rreg1[s_] := ParametricRegion[{Cos[t],`
`Sin[t], t}, {{t, 0, s}}]`

In[·]:= `RegionMeasure[rreg1[Pi]]`
Out[·]:= `Sqrt[2]*Pi`

In[·]:= `RegionMeasure[rreg1[Pi], 1]`
Out[·]:= `Sqrt[2]*Pi`

The two-dimensional measure is zero

In[·]:= `RegionMeasure[rreg1[Pi], 2]`
Out[·]:= `0`

and for dimension zero we have infinity:

In[·]:= `RegionMeasure[rreg1[Pi], 0]`
Out[·]:= `Infinity`

4.3.5 Unit (arclength) parametrization

In many textbooks on differential geometry, various formulas are given for curves with the so-called arclength parametrization. It is not easy to find it since one needs to solve complicated equations. Let us illustrate the process on the following example:

In[·]:= `curve[t_] := {t^2, t^3}`

In[·]:= `plotcurve = ParametricPlot[curve[t],`
`    {t, 0, 1}, PlotStyle -> {Opacity[0.5], Red}]`

Out[·]=

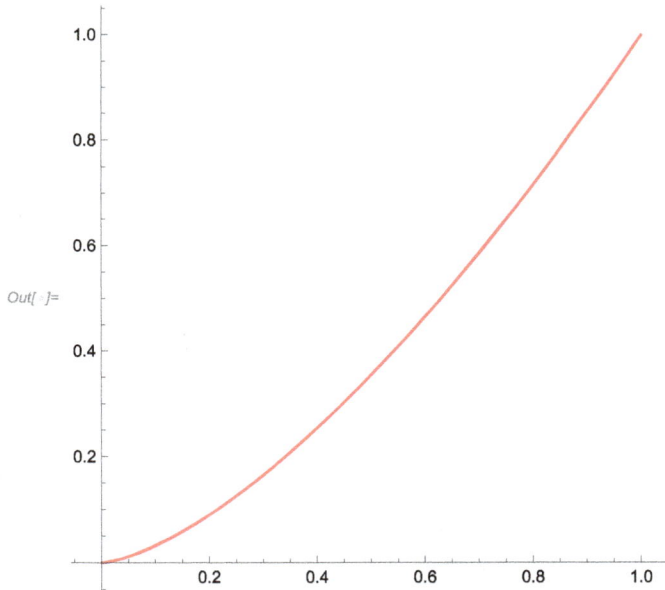

Figure 4.53

In[·]:= `ArcLength[curve[t], {t, 0, 1}] // N`
Out[·]:= `1.43971`

```
In[·]:= N[Integrate[Norm[D[curve[t], t]],
        {t, 0, 1}]]
Out[·]:= 1.43971
```

```
In[·]:= ArcLength[curve[u], {u, 0, t}]
Out[·]:= (1/27)*(-8 + (4 + 9*t^2)^(3/2))
```

The same answer is returned if we integrate directly (we omit the output):

```
In[·]:= expr = Integrate[Norm[D[curve[u], {u}]], {
        u, 0, t}, Assumptions -> {t > 0}];
```

To find the unit parametrization, we need to solve the following algebraic equation:

```
In[·]:= rule = Simplify[Solve[s == expr, t, Reals]
        [[2]], Assumptions -> {s > 0}][[1]]
Out[·]:= t -> Sqrt[Root[-16*s - 27*s^2 +
        16*#1 + 36*#1^2 + 27*#1^3 & , 1]]
```

```
In[·]:= curve1[s_] = {t^2, t^3} /. rule
Out[·]:= {Root[-16*s - 27*s^2 + 16*#1 + 36*#1^2
        {Root[-16*s - 27*s^2 + 16*#1 + 36*#1^2
        16*#1 + 36*#1^2 + 27*#1^3 & , 1]^(3/2)
```

```
In[·]:= plotcurve1 = ParametricPlot[curve1[s],
        {s, 0, 1}, PlotStyle -> {Opacity[0.5], Green}]
```

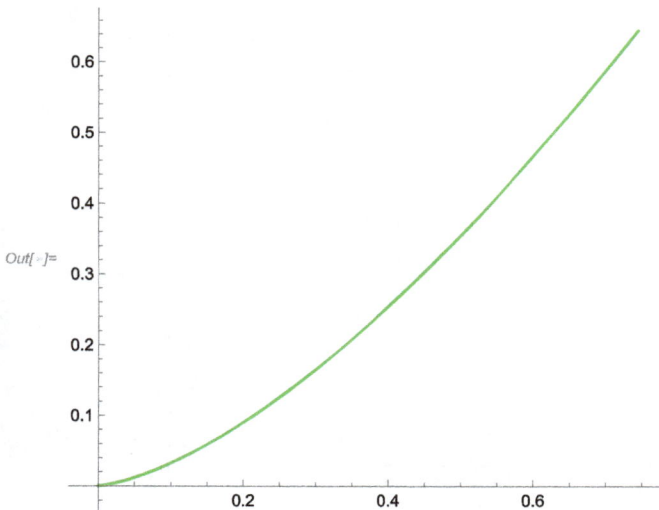

Figure 4.54

We can verify that we get the same curve:

In[·]:= Show[plotcurve, plotcurve1]

Out[·]=

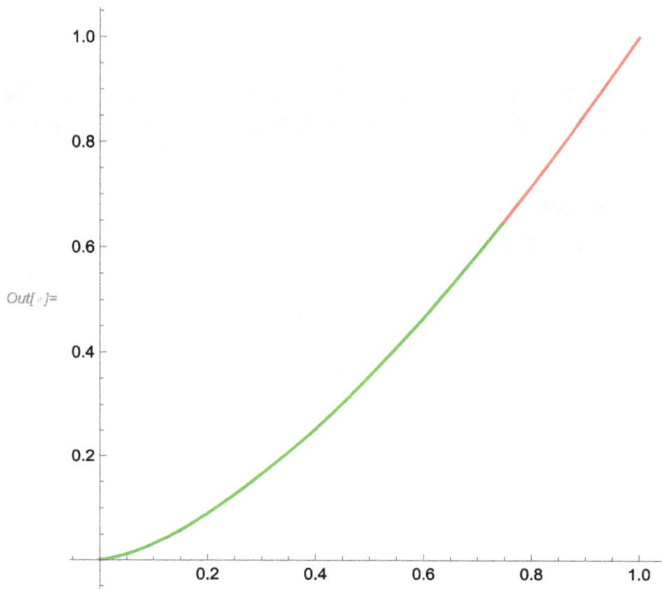

Figure 4.55

The method above can be used when **Mathematica**® can solve an equation. When there is no such solution, we can use another method with differential equations as follows. Since

$$\frac{ds}{dt} = \left\| \frac{df(t)}{dt} \right\|$$

and in our case

In[·]:= (Norm[D[curve[t], t]] /. Abs -> Identity)
        /. t -> t[s]
Out[·]:= Sqrt[4*t[s]^2 + 9*t[s]^4]

we have

$$\frac{dt(s)}{ds} = \frac{1}{\sqrt{4t(s)^2 + 9t(s)^4}}.$$

So we can solve the following differential equation to get the unit parametrization:

In[·]:= curvede[s_] = t[s] /. NDSolve[{D[t[s], s] ==
        1/Sqrt[4*t[s]^2 + 9*t[s]^4], t[0] == 0.01},
        t[s], {s, 0.01, 1}][[1]]

Out[ ]= **InterpolatingFunction** [⊞ ⟋ Domain: {{0.01, 1.}} | Output: scalar] [s]

Figure 4.56

The plot below gives a slightly shifted curve as we have a singularity at zero in the differential equation above and need to choose $t(0) = 0.01$ as our initial condition instead of $t(0) = 0$:

In[ ]:= ParametricPlot[{(curvede[s])^2,
        (curvede[s])^3}, {s, 0.01, 1}]

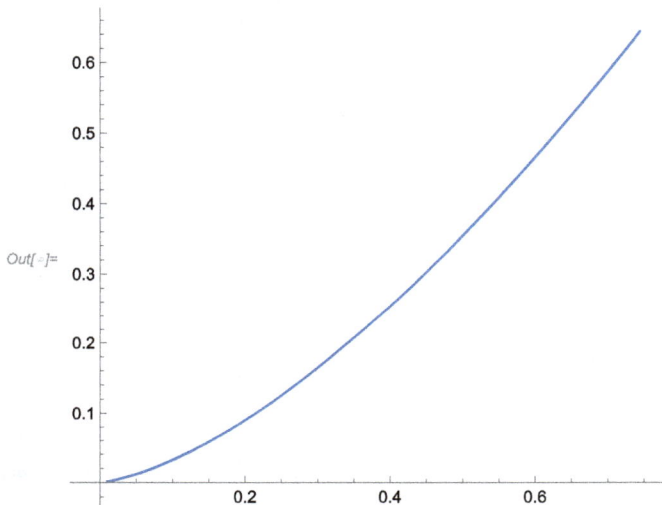

Figure 4.57

Another way to overcome the problem with the singularity at zero is to choose $t(0) = 1$ and shift the resulting curve by 1:

In[ ]:= curvede1[s_] = t[s] /. NDSolve[{D[t[s], s]
        == 1/Sqrt[4*t[s]^2 + 9*t[s]^4],
        t[0] == 1}, t[s], {s, 0, 10}][[1]]

Out[ ]= **InterpolatingFunction** [⊞ ⟋ Domain: {{0., 10.}} | Output: scalar] [s]

Figure 4.58

In[ ]:= plotcurve2 = ParametricPlot[{(curvede1[s] -
        1)^2, (curvede1[s] - 1)^3}, {s, 0, 10}]

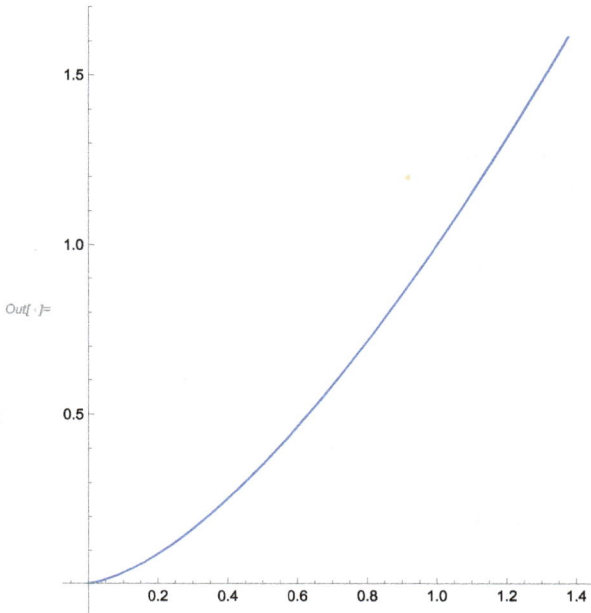

Figure 4.59

We see that the curve is the same:

In[·]:= Show[plotcurve, plotcurve2]

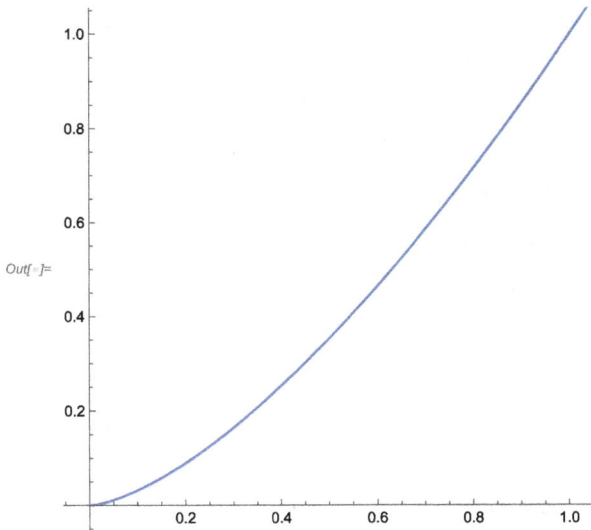

Figure 4.60

Note that if we want to calculate curvede1[11] with s outside of the interval $s \in [0, 10]$, extrapolation will be used:

> *In[·]:=* curvede1[11]
>
>   ···InterpolatingFunction: Input value {11} lies outside the range of data in
> the interpolating function. Extrapolation will be used.
> *Out[·]:=* 2.23814

For some built-in curves, the arclength function is available:

> *In[·]:=* Entity["PlaneCurve", "Circle"]["ArcLength"][a]
> *Out[·]:=* 2*a*Pi

> *In[·]:=* Entity["PlaneCurve", "Circle"]
>   ["ArcLengthFunction"][a][t]
> *Out[·]:=* a t

> *In[·]:=* Entity["PlaneCurve", "Ellipse"]
>   ["ArcLengthFunction"][2, 1][t]
> *Out[·]:=* EllipticE[t, -3]

> *In[·]:=* Entity["PlaneCurve", "BulletNoseCurve"]
>   ["ArcLengthFunction"]
> *Out[·]:=* Missing["NotAvailable"]

4.3.6 Curvature of a curve in $\mathbb{R}^n$

Let $\alpha : I \to \mathbb{R}^n$ be a smooth curve. As mentioned above, sometimes it is convenient to introduce our own definitions, even if there are built-in functions which perform the same task. In this section we shall use the norm defined by

> *In[·]:=* norm[(v_)?VectorQ] := Sqrt[v . v]

We shall also define our own function normalize by

> *In[·]:=* normalize[(v_)?VectorQ] := v/norm[v]

The unit tangent vector field to α is defined by

> *In[·]:=* uTangent[(alpha_)?VectorQ, t_] :=
>   normalize[D[alpha, t]]

The curvature of α at the point t is defined as the "rate of turning" of uTangent, i. e., $\|(d(\text{uTangent})/dt)/(d\alpha/dt)\|$:

> *In[·]:=* k[alpha_, t_] := Sqrt[D[uTangent[
>   alpha, t], t] . D[uTangent[alpha, t], t]/
>   D[alpha, t] . D[alpha, t]]

In[·]:= Simplify[k[{t, t}, t], Element[_, Reals]]
Out[·]:= 0

The quickest way is to use the built-in function FrenetSerretSystem (which appeared in version 10 of **Mathematica**®). In two dimensions it returns the (signed) curvature and a pair of vectors, namely the unit tangent vector and the unit normal perpendicular to it:

In[·]:= FrenetSerretSystem[{3 t, 4 t}, t]
Out[·]:= {{0}, {{3/5, 4/5}, {-(4/5), 3/5}}}

This function can calculate these quantities in other coordinate systems, for instance,

In[·]:= FrenetSerretSystem[{1, 0, -4 t}, t,
        "Cylindrical"]
Out[·]:= {{0, 0}, {{0, 0, -1}, {0, 0, 0}, {0, 0, 0}}}

If we write out the curvature formula explicitly, computation will be faster. It is convenient to substitute α as a function of t in the following formula:

In[·]:= k1[alp_, t_Symbol] := Sqrt[D[alp, t] .
        D[alp, t]*D[alp, {t, 2}] . D[alp, {t, 2}]
        - (D[alp, t] . D[alp, {t, 2}])^2]/
        (D[alp, t] . D[alp, t])^(3/2)

In[·]:= Simplify[k1[{t, t}, t], Element[_, Reals]]
Out[·]:= 0

For a curve given as a function, we can use another definition:

In[·]:= k2[alp_][t_] := Sqrt[Derivative[1][alp][t]
        . Derivative[1][alp][t]*Derivative[2][alp][t]
        . Derivative[2][alp][t] - (Derivative[1][alp][t] .
        Derivative[2][alp][t])^2]/(Derivative[1][alp][t]
        . Derivative[1][alp][t])^(3/2)

In[·]:= circle[t_] := {Cos[t], Sin[t]}

In[·]:= k2[circle][t] // FullSimplify
Out[·]:= 1

In this way we can directly calculate the value of the curvature at a given point:

In[·]:= k2[circle][0]
Out[·]:= 1

Note that we could have used the same name k in the definition of k1 and k2, but we want to distinguish them in the following example of a curve with singularities to emphasize which definition is used.

In[·]:= hypotrochoid[a_, b_, h_][t_] :=
    . {h*Cos[(t*(a - b))/b] + (a - b)*Cos[t],
    (a - b)*Sin[t] - h*Sin[(t*(a - b))/b]}

In[·]:= curveht = ParametricPlot[
    hypotrochoid[5, 1, 2][t], {t, 0, 2*Pi}]

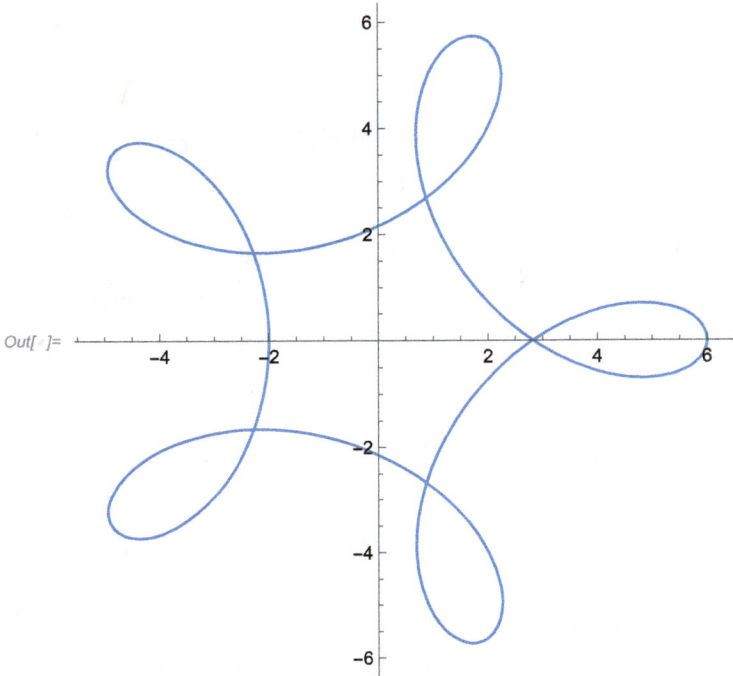

Figure 4.61

In[·]:= k1[hypotrochoid[5, 1, 2][s], s] /. s -> 0.1
Out[·]:= 1.3385

In[·]:= k2[hypotrochoid[5, 1, 2]][0.1]
Out[·]:= 1.3385

In fact, in two dimensions the curvature is signed, and we shall discuss this in the following section:

In[·]:= FrenetSerretSystem[hypotrochoid[
    5, 1, 2][t], t][[1, 1]] /. t -> 0.1
Out[·]:= -1.3385

In the following interactive illustration, we plot the hypotrochoid and the absolute value of the curvature of a point moving on it.

```
In[·]:= Manipulate[Grid[{{Show[Graphics[{Red,
        PointSize[0.02], Point[hypotrochoid[
        5, 1, 2][t]]}], curveht], Style[
        k2[hypotrochoid[5, 1, 2]][t], Red]}}],
        {t, 0., 2*Pi, Appearance -> "Labeled"}]
```

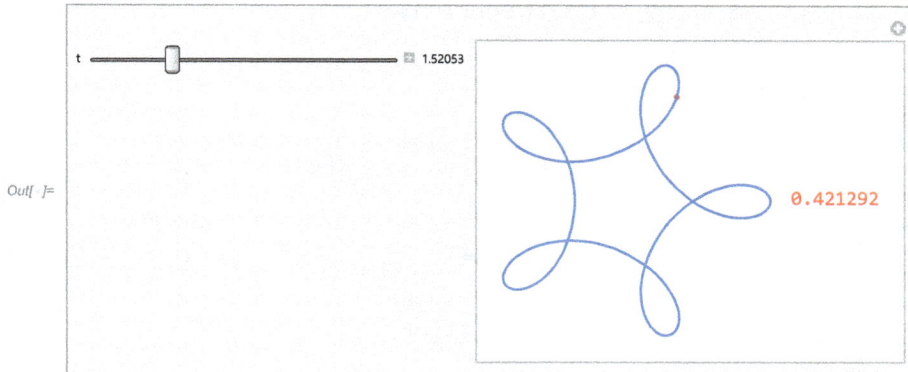

Figure 4.62

4.3.7 Signed curvature of a plane curve

The general notion of curvature of a curve in $\mathbb{R}^n$ does not take account of a sign. For curves in $\mathbb{R}^2$, there is a more refined notion of signed curvature that does. It is defined by

```
In[·]:= ks[alp_][t_] := Derivative[2][alp][t]
        . RotationTransform[Pi/2][Derivative[1][alp][t]]/
        (Derivative[1][alp][t] . Derivative[1][alp][t])^(3/2)
```

Here α must be a 2D vector-valued function and t must be a real number. Now we get the same answer as above in the previous subsection when we used FrenetSerret-System:

```
In[·]:= ks[hypotrochoid[5, 1, 2]][0.1]
Out[·]:= -1.3385
```

The function RotationTransform is one of **Mathematica**®'s geometric functions which rotates a vector through a given angle ($\pi/2$ above).

```
In[·]:= alp[t_] := {t, -t}
```

In[·]:= ks[alp][t]
Out[·]:= 0

The above definition works for functions $\mathbb{R} \to \mathbb{R}^2$. We give another one for a pair of functions:

In[·]:= ks1[{x_, y_}][t_] = ((-Derivative[1]
        [y][t])*Derivative[2][x][t] +
        Derivative[1][x][t]*Derivative[2][y][t])/
        (Derivative[1][x][t]^2 + Derivative[1]
        [y][t]^2)^(3/2);

and one more for a pair of expressions:

In[·]:= ks2[{x_, y_}, t_] := ks1[{Function[
        x /. t -> #], Function[y /. t -> #]}][t]

The positive function $1/|\kappa_s(\alpha)|$ is called the radius of curvature of the curve α.

Next we calculate the signed curvature of a circle given in a vector form as a pair of expressions:

In[·]:= ks2[{Cos[t], Sin[t]}, t] // Simplify
Out[·]:= 1

Here is the same thing for a pair of functions:

In[·]:= ks1[{Cos , Sin}][t] // Simplify
Out[·]:= 1

and here for a single function with values in $\mathbb{R}^2$:

In[·]:= ks[{Cos[#], Sin[#]} &][t] // Simplify
Out[·]:= 1

For our second example, we consider the bullet nose curve. Mathematica® knows only the algebraic equation of this curve:

In[·]:= Entity["PlaneCurve", "BulletNoseCurve"]
        ["AlgebraicEquation"]

Out[·]= $\text{Function}\left[\{a, b\}, \text{Function}\left[\{x, y\}, -b^2 x^2 + a^2 y^2 - x^2 y^2\right]\right]$

Figure 4.63

In[·]:= Entity["PlaneCurve", "BulletNoseCurve"]
        ["ParametricEquations"]
Out[·]:= Missing["NotAvailable"]

But we can verify that the following gives a parametization of the bullet nose curve:

In[·]:= `TrigExpand[(-b^2)*x^2 + a^2*y^2 - x^2*y^2`
`           /. {x -> a*Cos[t], y -> b*Cot[t]}]`
Out[·]:= 0

So we shall use the following parametization:

In[·]:= `bulletnose[a_, b_][t_] := {a Cos[t], b Cot[t]}`

We can calculate the curvature using our function:

In[·]:= `ksbn = ks[bulletnose[a, b]][Pi/3]`
Out[·]:= `-((2*a*b)/((3*a^2)/4 + (16*b^2)/9)^(3/2))`

We can use the built-in information for the bullet nose curve and compare the results. We first find the coordinates of the point corresponding to $t = \pi/3$:

In[·]:= `{a Cos[t], b Cot[t]} /. t -> Pi/3`
Out[·]:= `{a/2, b/Sqrt[3]}`

We can check the built-in implicit curvature at this point:

In[·]:= `ksbn1 = FullSimplify[Entity[`
`         "PlaneCurve", "BulletNoseCurve"]`
`         ["ImplicitCurvature"][a, b][a/2,`
`         b/Sqrt[3]]]`
Out[·]:= `-((432*Sqrt[27*a^4*b^2 + 64*a^2*b^4])/`
`         (27*a^2 + 64*b^2)^2)`

We have obtained a slightly different expression, but we can easily check that the result is the same as above:

In[·]:= `ksbn - ksbn1 // Together // ToRadicals`
`         // PowerExpand // Simplify`
Out[·]:= 0

4.3.8 Evolutes and involutes for plane curves

Let us define the circle of curvature. A point $p \in \mathbb{R}^2$ is called the center of curvature at q of a curve $\alpha : (a, b) \to \mathbb{R}^2$ provided there is a circle y with center p which is tangent to α at q and such that the curvatures of α and y are the same at q. The centers of the circles of curvature of a curve α form a curve called the evolute of α. The equation of the evolute is $\alpha(t) + (1/\kappa_s(t))N(t)$, where N is the unit normal vector.

```
In[·]:= evolute[f_][t_] := Simplify[
        RotationTransform[Pi/2][Derivative[1]
        [f][t]]/(Sqrt[Derivative[1][f][t]
        . Derivative[1][f][t]]*ks[f][t]) + f[t]]
```

The unit normal vector can also be obtained from FrenetSerretSystem.

```
In[·]:= curvealp[t_] := {x[t], y[t]}
```

Let us compare our definition with that obtained by using FrenetSerretSystem:

```
In[·]:= Simplify[evolute[curvealp][t] -
        ({x[t], y[t]} + FrenetSerretSystem[
        {x[t], y[t]}, t][[2, 2]]/
        FrenetSerretSystem[{x[t], y[t]},
        t][[1, 1]])]
Out[·]:= {0, 0}
```

As earlier, we define the evolute for curves given by pairs of expressions and pairs of functions. For simplicity we use the same name and do not distinguish as done before with ks, ks1, ks2.

```
In[·]:= evolute[{x_, y_}][t_] =
        evolute[curvealp][t];
```

```
In[·]:= evolute[{x_, y_}, t_] := evolute[{
        Function[x /. t -> #], Function[
        y /. t -> #]}][t]
```

```
In[·]:= Simplify[evolute[{4*Sin[#1] & ,
        2*Cos[#1] & }][t]]
Out[·]:= {3*Sin[t]^3, -6*Cos[t]^3}
```

```
In[·]:= evolute[{4*Sin[#1], 2*Cos[#1]} & ][t]
Out[·]:= {3*Sin[t]^3, -6*Cos[t]^3}
```

```
In[·]:= grevol = ParametricPlot[{{
        4*Cos[t], 2*Sin[t]}, evolute[{
        4*Cos[#1] & , 2*Sin[#1] & }][t]},
        {t, 0, 2*Pi}]
```

Out[ ]=

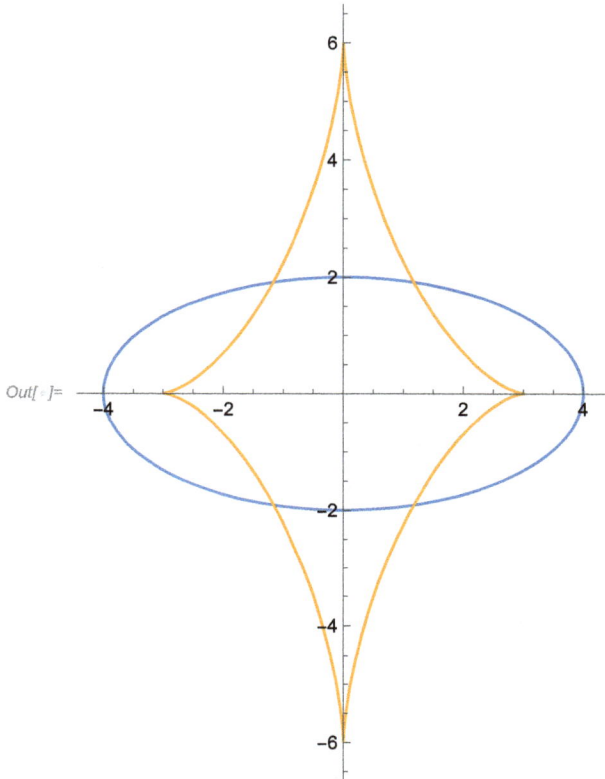

Figure 4.64

In the following dynamic illustration, we can see the red point which is a center of curvature, the curve (blue), the evolute (orange), and the circle of curvature (red).

```
In[·]:= Manipulate[Show[grevol, Graphics
       [{Red, PointSize[0.02]}, Point[evolute
       [{4*Cos[#1] & , 2*Sin[#1] & }][t]],
       Circle[evolute[{4*Cos[#1] & ,
       2*Sin[#1] & }][t], 1/Abs[ks1[{
       4*Cos[#1] & , 2*Sin[#1] & }][t]]]}],
       PlotRange -> 8], {t, 0, 2*Pi}]
```

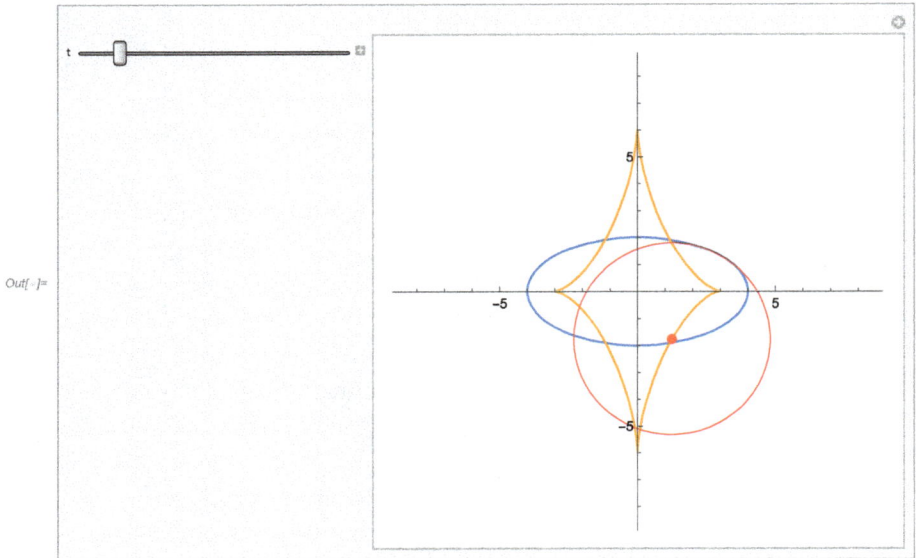

Figure 4.65

The involute is the operation of taking the inverse to the operation that assigns to a curve its evolute. The relation between evolute and involute is analogous to the relation between differentiation and integration. Thus, the involute essentially means the initial curve for a given evolute. The definition of involute uses integration but this very rarely can be successfully done symbolically. Therefore we will mostly use a numerical definition which is suitable for finding involutes graphically.

We can first give a symbolic definition of an involute. In those cases when involutes can be found explicitly, drawing them becomes much quicker. To define an involute, we need to give not only a curve but also a starting value c of the parameter.

```
In[·]:= involute[alp_, c_][t_] :=
        Assuming[Element[t, Reals],
        Simplify[(-(1/Sqrt[Derivative[1]
        [alp][t] . Derivative[1][alp][t]]))
        *Derivative[1][alp][t]*Integrate[
        Sqrt[Derivative[1][alp][s] .
        Derivative[1][alp][s]], {s, c, t},
        Assumptions -> {t >= c >= 0}] +
        alp[t]]]]
```

```
In[·]:= involute[{x_, y_}, c_, t_] :=
        involute[Function[{x, y} /.
        t -> #], c][t]
```

The formula for the involute for the simple curve below needs elliptic functions:

In[·]:= inv[t_] = involute[{2 Cos[#],
    3 Sin[#]} &, 0][t]

Out[·]= $\left\{2\,\mathrm{Cos}\,[\,t\,] + \dfrac{6\,\left(\mathrm{EllipticE}\left[\pi\,\mathrm{FractionalPart}\left[\frac{t}{\pi}\right],\,\frac{5}{9}\right] + 2\,\mathrm{EllipticE}\left[\frac{5}{9}\right]\,\mathrm{IntegerPart}\left[\frac{t}{\pi}\right]\right)\,\mathrm{Sin}\,[\,t\,]}{\sqrt{9\,\mathrm{Cos}\,[\,t\,]^2 + 4\,\mathrm{Sin}\,[\,t\,]^2}},\right.$

$\left.3\,\mathrm{Sin}\,[\,t\,] - \dfrac{9\,\mathrm{Cos}\,[\,t\,]\,\left(\mathrm{EllipticE}\left[\pi\,\mathrm{FractionalPart}\left[\frac{t}{\pi}\right],\,\frac{5}{9}\right] + 2\,\mathrm{EllipticE}\left[\frac{5}{9}\right]\,\mathrm{IntegerPart}\left[\frac{t}{\pi}\right]\right)}{\sqrt{9\,\mathrm{Cos}\,[\,t\,]^2 + 4\,\mathrm{Sin}\,[\,t\,]^2}}\right\}$

Figure 4.66

In[·]:= ParametricPlot[inv[t], {t, 0, 8 Pi}]

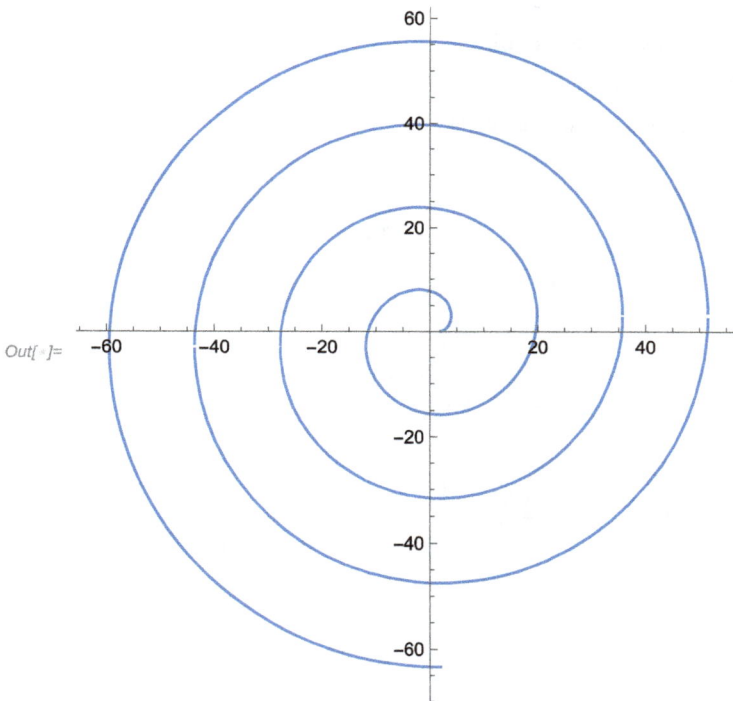

Out[·]=

Figure 4.67

We can check whether **Mathematica**® has this involute built-in:

In[·]:= Entity["PlaneCurve", "Ellipse"]
    ["Involute"]

Out[ ]= ellipse involute

Figure 4.68

We see that Mathematica® returns a slightly different answer for parametric equations (which agrees numerically with ours):

In[·]:= Entity["PlaneCurve", "EllipseInvolute"]
        ["ParametricEquations"][2, 3][t]

$$Out[·]= \left\{2\left(\text{Cos}[t] + \frac{3\sqrt{2}\ \text{EllipticE}\left[t, \frac{5}{9}\right]\ \text{Sin}[t]}{\sqrt{13 + 5\ \text{Cos}[2\,t]}}\right), 3\left(-\frac{3\sqrt{2}\ \text{Cos}[t]\ \text{EllipticE}\left[t, \frac{5}{9}\right]}{\sqrt{13 + 5\ \text{Cos}[2\,t]}} + \text{Sin}[t]\right)\right\}$$

Figure 4.69

The built-in graphics object is also similar:

In[·]:= Entity["PlaneCurve",
        "EllipseInvolute"]["Graphics"]

Out[ ]=

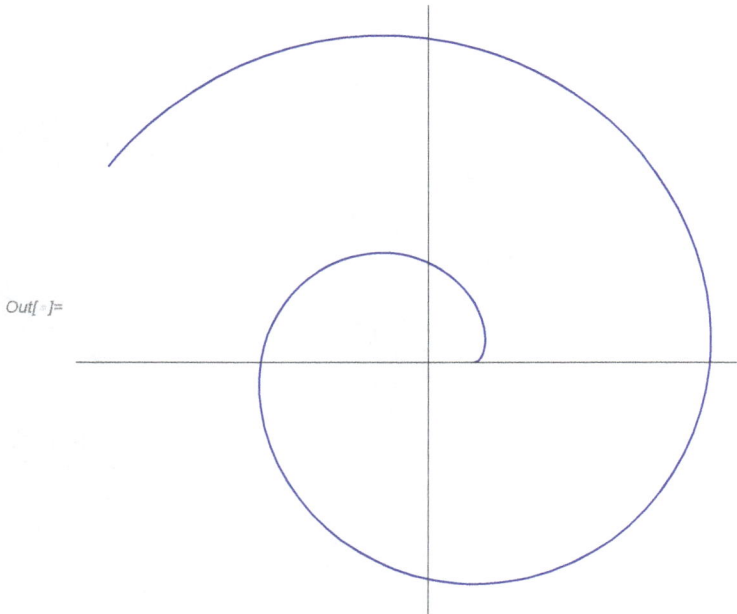

Figure 4.70

A safer and quicker way is to integrate numerically.

```
In[·]:= nInvolute[alp_, c_][t_] :=
        -(Derivative[1][alp][t]*NIntegrate[
        Norm[Derivative[1][alp][s]], {s, c, t}])
        /Norm[Derivative[1][alp][t]] +
        alp[t]
```

```
In[·]:= nInvolute[{Cos[#1], Sin[#1]} & , 0][0.3]
Out[·]:= {1.04399, 0.00891926}
```

Drawing the graph of an involute, even of a circle, takes a long time.

```
In[·]:= ParametricPlot[nInvolute[{Cos[#], Sin[#]}
        &, 0][t], {t, 0, 8 Pi}]
```

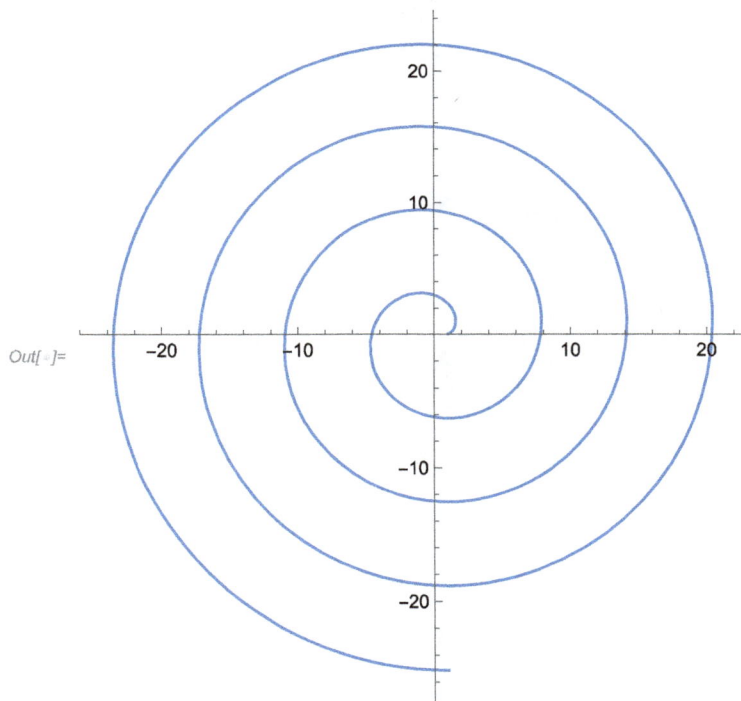

Figure 4.71

As mentioned before, the evolute of the involute returns the original curve:

```
In[·]:= evolute [involute[{x[t], y[t]},
         0, t], t] // FullSimplify
Out[·]:= {x[t], y[t]}
```

We can also define evolutes of evolutes and same for involutes. It is convenient to use NestList.

```
In[·]:= listevol = NestList[evolute[#, t] &,
         {2 Cos[t], 3 Sin[t]}, 4];
```

```
In[·]:= Length[listevol]
Out[·]:= 5
```

```
In[·]:= ParametricPlot[Evaluate[listevol],
         {t, 0.01, 2 Pi}]
```

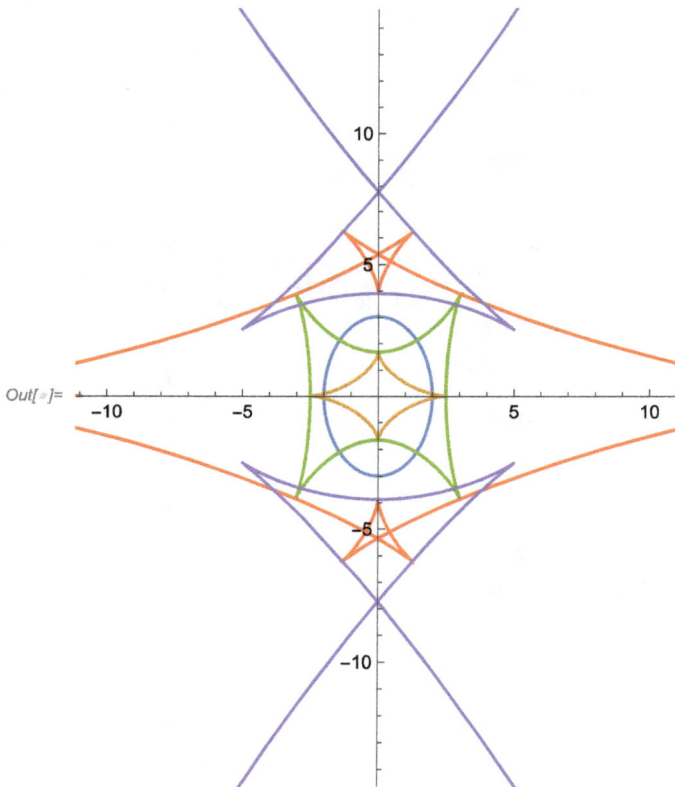

Figure 4.72

```
In[·]:= listinvol = NestList[involute[
        #, 0, t] &, {Cos[t], Sin[t]}, 2];
```

```
In[·]:= ParametricPlot[Evaluate[listinvol],
        {t, 0.1, 2 Pi}, PlotStyle ->
        {Red, Green, Blue}]
```

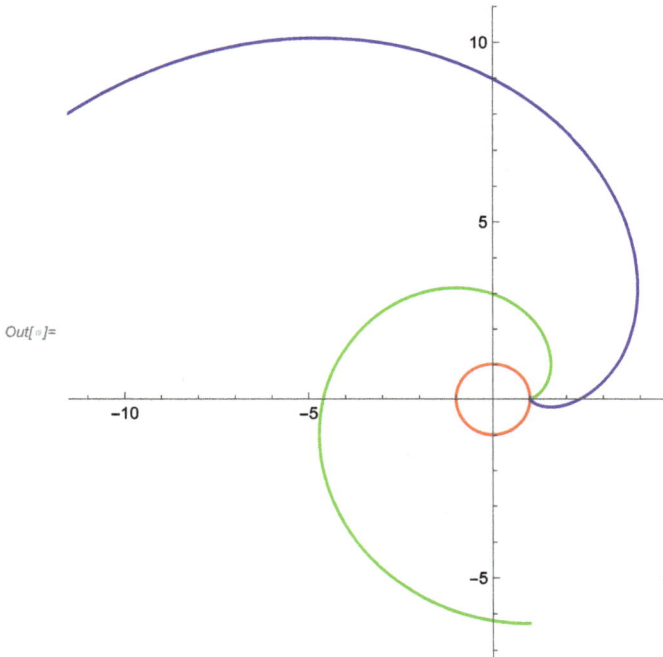

Out[·]=

Figure 4.73

Evolutes and involutes can also be defined similarly for space curves but the formulas
are much longer so we omit them.

4.3.9 Curvature of implicitly defined plane curves

Suppose we have a plane curve given by an equation $F(x, y) = 0$. Here is an algebraic
example:

```
In[·]:= F[x_, y_] := x^3 + 3*x*y - y^3 - 3
```

It is in general difficult to find intervals for x and y to see an interesting behavior of the
curve:

```
In[·]:= ContourPlot[F[x, y] == 0, {x, -5, 5},
        {y, -5, 5}]
```

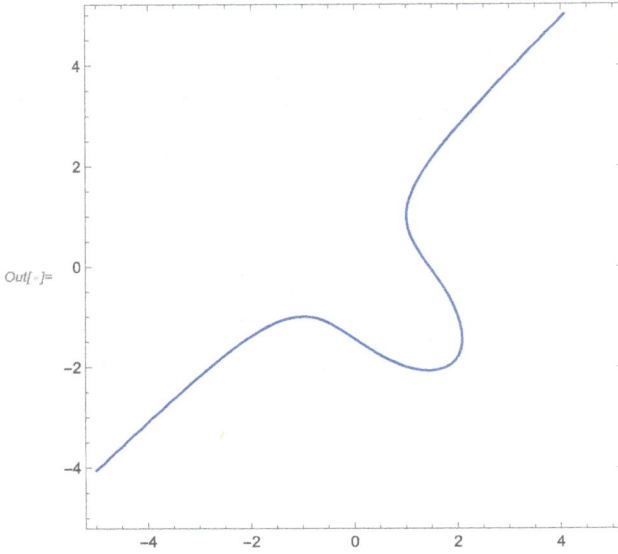

Figure 4.74

Here is a nonalgebraic example:

In[·]:= F1[x_, y_] := x Sin[y] + y Sin[x]

In[·]:= ContourPlot[F1[x, y] == 0, {x, -20, 20},
    {y, -20, 20}]

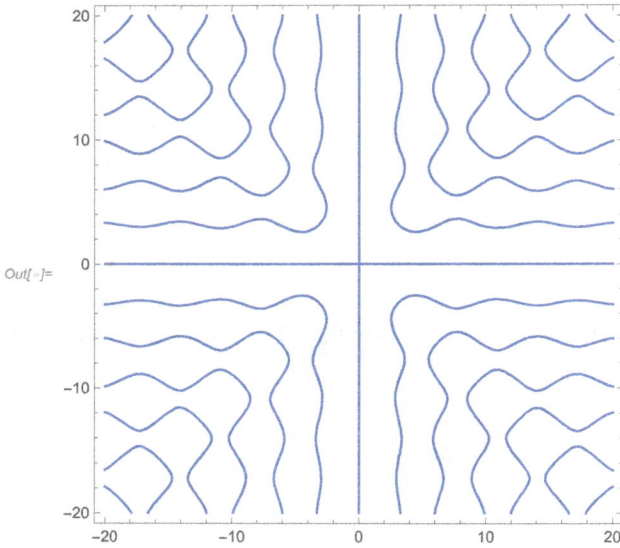

Figure 4.75

The question now is how to compute the curvature of a plane curve defined implicitly at a given point (a, b) lying on the curve. Finding a parametric representation locally and numerically is not difficult in general. The crucial observation is that we can find the Frenet–Serret system without the need for a parametric representation. It is very easy to find a unit normal and unit tangent fields. To calculate the normal vector, we use Grad:

```
In[·]:= normal[x_, y_] = Grad[F[x, y], {x, y}]
Out[·]:= {3*x^2 + 3*y, 3*x - 3*y^2}
```

```
In[·]:= Show[{ContourPlot[F[x, y] == 0,
         {x, -5, 5}, {y, -5, 5}], Graphics[
         Arrow[{{1, -2}, {1, -2} + Normalize[
         normal[1, -2]]}]], VectorPlot[{
         3*x^2 + 3*y, 3*x - 3*y^2}, {x, -5, 5},
         {y, -5, 5}, VectorColorFunction ->
         (Red & ), VectorPoints -> 30]}]
```

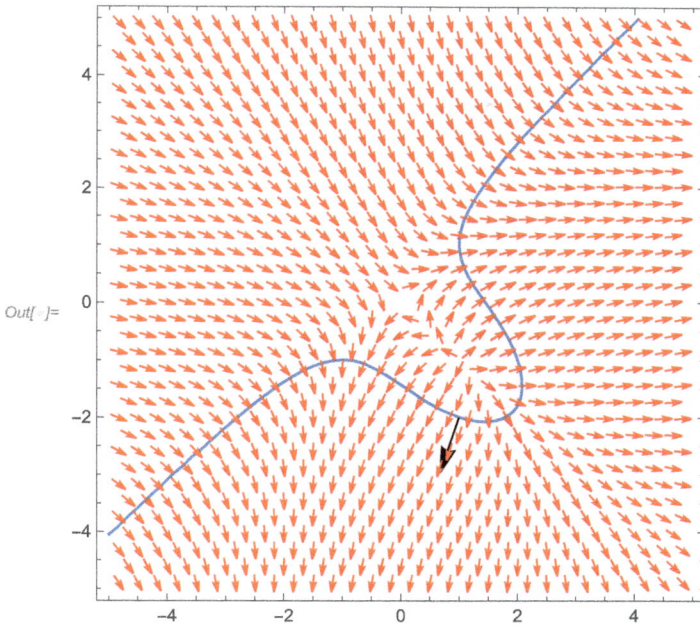

Figure 4.76

Note that Mathematica® also has the functions StreamPlot which was updated in version 12.2:

```
In[·]:= StreamPlot[{3*x^2 + 3*y, 3*x - 3*y^2},
         {x, -5, 5}, {y, -5, 5}]
```

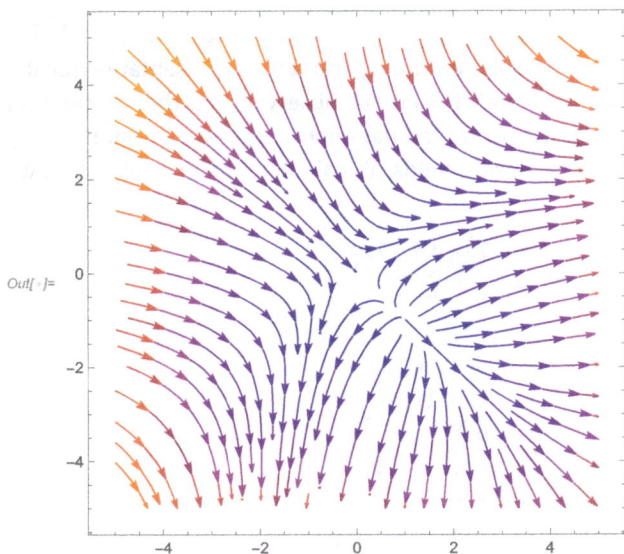

Figure 4.77

The function StreamPlot3D was introduced in version 12.3:

In[ ]:= StreamPlot3D[{x, y, z}, {x, -1, 1},
  {y, -1, 1}, {z, -1, 1}]

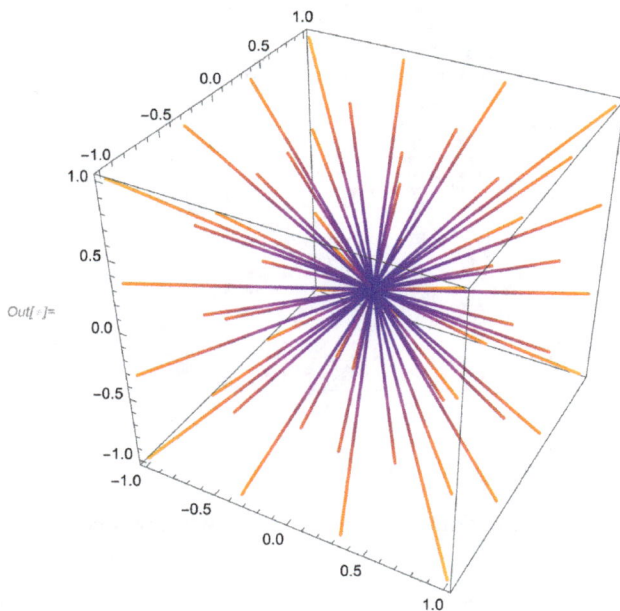

Figure 4.78

Compare with

In[·]:= VectorPlot3D[{x, y, z}, {x, -1, 1},
{y, -1, 1}, {z, -1, 1}]

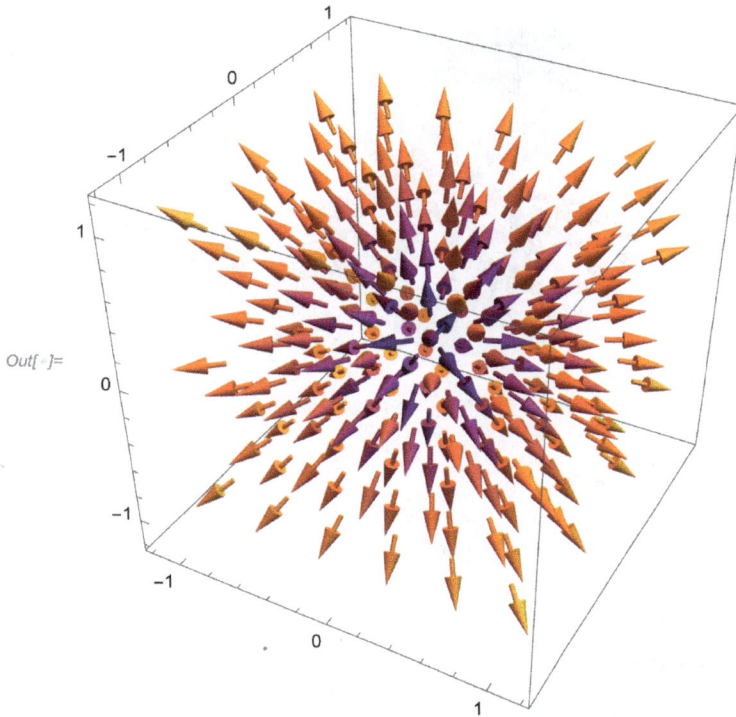

Out[·]=

Figure 4.79

In[·]:= VectorPlot3D[{x, y, z},
Element[{x, y, z}, Ball[]]]

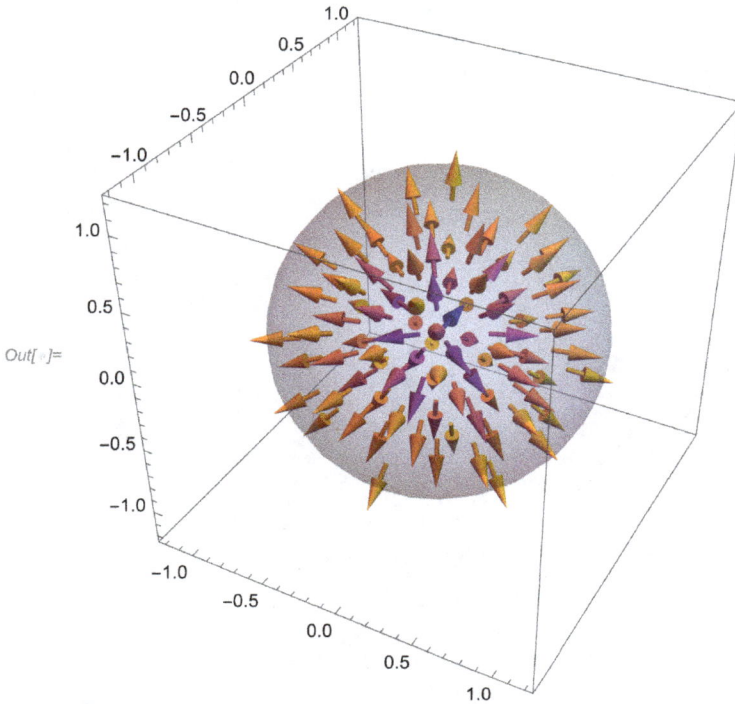

Figure 4.80

The tangent vector is defined by

```
In[·]:= tangent[x_, y_] :=
         RotationTransform[Pi/2]
         [normal[x, y]]
```

```
In[·]:= tangent[x, y]
Out[·]:= {-3*x + 3*y^2, 3*x^2 + 3*y}
```

```
In[·]:= Show[{ContourPlot[F[x, y] == 0,
         {x, -5, 5}, {y, -5, 5}],
         Graphics[Arrow[{{1, -2}, {1, -2} +
         Normalize[tangent[1, -2]]}]]}]
```

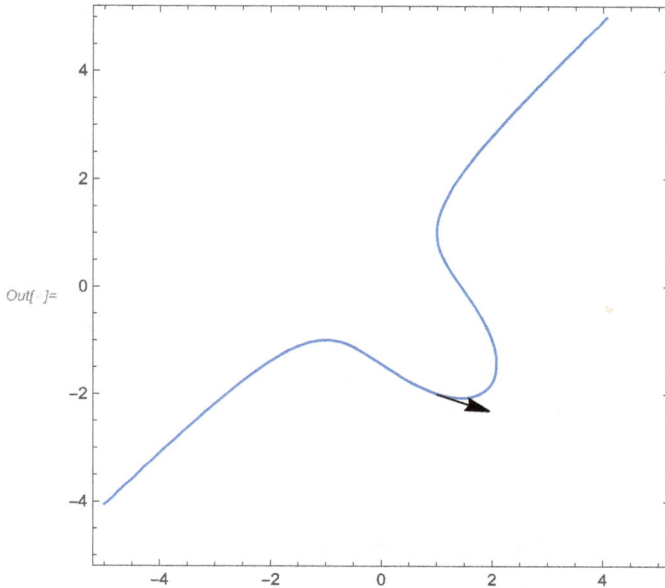

Figure 4.81

Now we can obtain the formula for curvature using the same method as for a parametric description:

```
In[·]:= kimp[f_, {x_, y_}] := -((-(2*
         D[f, {x}]*D[f, {y}]*D[f, {x}, {y}])
         + D[f, {x, 2}]*D[f, {y}]^2 +
         D[f, {x}]^2*D[f, {y, 2}])/
         (D[f, {x}]^2 + D[f, {y}]^2)^(3/2))
```

```
In[·]:= kimp[F[x, y], {x, y}] /. {x -> 1, y -> 1}
Out[·]:= 1
```

Recall that for a parametric plane curve we have the following formula for curvature:

```
In[·]:= ks[{x_, y_}][t_] = ((
         -Derivative[1][y][t])*Derivative[2][x]
         t] + Derivative[1][x][t]*Derivative[2]
         [[y][t])/(Derivative[1][x][t]^2 +
         Derivative[1][y][t]^2)^(3/2);
```

We can prove that the formulas are equivalent by differentiating twice the curve equation $F(x(t), y(t)) = 0$.

```
In[·]:= subst1 = Solve[D[F[x[t], y[t]] ==
         0, t], Derivative[1][y][t]][[1]];
```

```
In[·]:= subst2 = Solve[Simplify[D[F[x[t],
        y[t]] == 0, {t, 2}] /. subst1],
        Derivative[2][y][t]][[1]];
```

```
In[·]:= (ks[{x, y}][t] //. subst2 //. subst1
        // Together) // PowerExpand;
```

```
In[·]:= % - (kimp[F[x, y], {x, y}] /.
        {x -> x[t], y -> y[t]}) // Together
```
```
Out[·]:= 0
```

One can also solve the problem numerically (using the Inverse Function Theorem) by constructing an approximate local parametrization provided that the curve is nonsingular. Assume that we want to find y as a function of x. We shall use Mathematica®'s ability to solve ordinary differential equations numerically.

```
In[·]:= eq = D[F[x, y[x]], x];
```

We need to find some point on the curve, e. g., let us take $x = 0$:

```
In[·]:= point = First[y /. Solve[F[0, y] == 0, y,
        Reals]]
```
```
Out[·]=  ⊙ -1.44...
```

Figure 4.82

```
In[·]:= solcurve = y /. NDSolve[{eq == 0, y[0] ==
        point}, y, {x, -0.5, 0.5}, WorkingPrecision
        -> 20][[1]]
```
```
Out[·]= InterpolatingFunction[ ⊞ ⩘ Domain: {{-0.50000000000000000000, 0.50000000000000000000}}
                                      Output: scalar ]
```

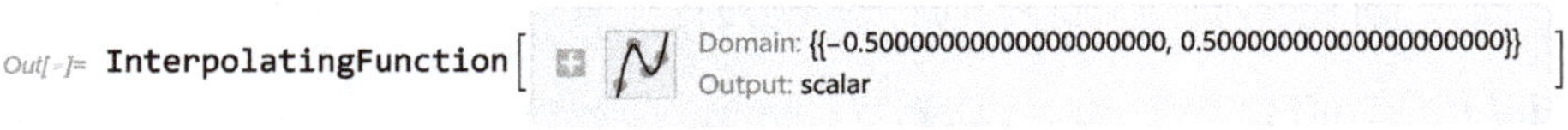

Figure 4.83

```
In[·]:= solcurve[0]
```
```
Out[·]:= -1.4422495703074083823
```

```
In[·]:= F[0, solcurve[0]] // N
```
```
Out[·]:= 0.
```

We can see that the solution gives us a piece of the curve, so we indeed found a parametric representation of our curve.

```
In[·]:= grsolcurve = Plot[solcurve[x], {x,
        -0.5, 0.5}, PlotStyle -> {Green,
        Dashing[0.2]}];
```

In[·]:= curveF = ContourPlot[F[x, y] == 0,
    {x, -0.5, 0.5}, {y, point - 1, point + 1},
    Axes -> True, ColorFunction -> Function
    [{x, y}, Red]];

In[·]:= Show[curveF, grsolcurve]

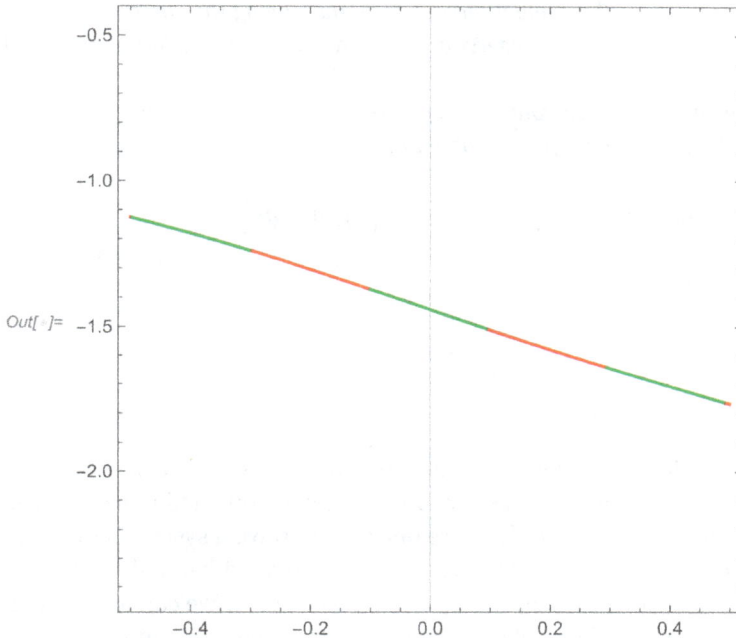

Figure 4.84

We can now compare the curvatures computed exactly and numerically using our parametrization and see that around this chosen point the curve is like a straight line:

In[·]:= kimp[F[x, y], {x, y}] /. {x -> 0,
    y -> point} // N
Out[·]:= 0.

In[·]:= Abs[ks[{# &, solcurve[#] &}][0]]
Out[·]:= 7.04*10^(-6)

Note that an implicit equation of a curve does not determine its orientation, hence we can only compute unsigned curvature. Also, when using this method, that is, representing a function given implicitly by an explicit interpolation function obtained by solving a differential equation numerically, errors in numerical differentiation may make the answer inaccurate.

4.3.10 Drawing plane curves with assigned curvature

In this section we shall explain the Fundamental Theorem for Plane Curves which gives the way to express the curve in terms of its curvature.

Theorem 1 ([1, Theorems 5.13 and 5.14]). *(Uniqueness) Let α and γ be unit-speed regular plane curves defined on the same interval (a, b) and having the same signed curvature. Then there is an orientation-preserving Euclidean motion (translation, rotation) mapping α into γ.*

(Existence) Given a piecewise-continuous function κ : (a, b) → ℝ, the unit curve β on this interval with curvature κ is parametrized by

$$\beta(s) = \left(c + \int \cos \theta(s)\, ds,\ d + \int \sin \theta(s)\, ds \right),$$

where

$$\theta(s) = \int \kappa(s)\, ds + \theta_0$$

and c, d, θ_0 are constants.

There exists a similar Fundamental Theorem for Space Curves. A very similar program that draws curves in space with assigned positive curvature and torsion can be easily written, but the code is longer [1] since one should solve a system of ordinary differential equations coming from the Serret–Frenet formulas with initial conditions. We shall therefore consider a simpler example of a plane curve. We can use the theorem above to find (numerically) a plane curve with any given curvature. Below we shall use the function NDSolve to solve a system of differential equations with initial conditions:

```
In[·]:= fundth[f_, a_ : 0, {c_ : 0, d_ : 0,
       theta_ : 0}, {smin_ : -10, smax_ : 10}] :=
       Module[{x, y, th}, Flatten[{x, y}
       /. NDSolve[{Derivative[1][x][ss] ==
       Cos[th[ss]], Derivative[1][y][ss] ==
       Sin[th[ss]], Derivative[1][th][ss] ==
       f[ss], x[a] == c, y[a] == d,
       th[a] == theta}, {x, y, th},
       {ss, smin, smax}]]]
```

Here is a curve whose curvature is $k(s) = s + \sin(s)$:

In[ ]:= `cur[t_] := Through[fundth[(# + Sin[#])`
`& , 0, {0, 0, 0}, {-18, 18}][t]]`

In[ ]:= `ParametricPlot[cur[t], {t, -10, 10},`
`PlotPoints -> 10]`

Out[ ]=

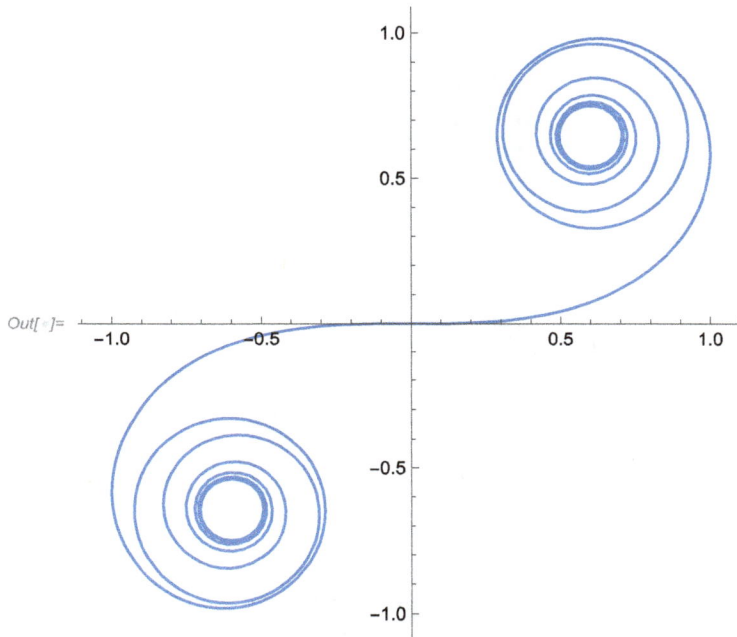

Figure 4.85

4.3.11 Envelopes

Envelopes are considered in detail in [1], see also [19]. Let us consider a function F of two variables x and y, and one parameter t. We can think of $F_t(x, y) = F(x, y, t) = 0$ as a family of plane curves parametrized by the parameter t.

In[ ]:= `F[x_, y_, t_] := (x - t)^2 + (y - t^2)^2 - 4`

If we consider F as a function of three variables, then we get a surface in $\mathbb{R}^3$.

In[ ]:= `ContourPlot3D[F[x, y, t] == 0, {x,`
`-12, 12}, {y, -12, 12}, {t, -5, 5}]`

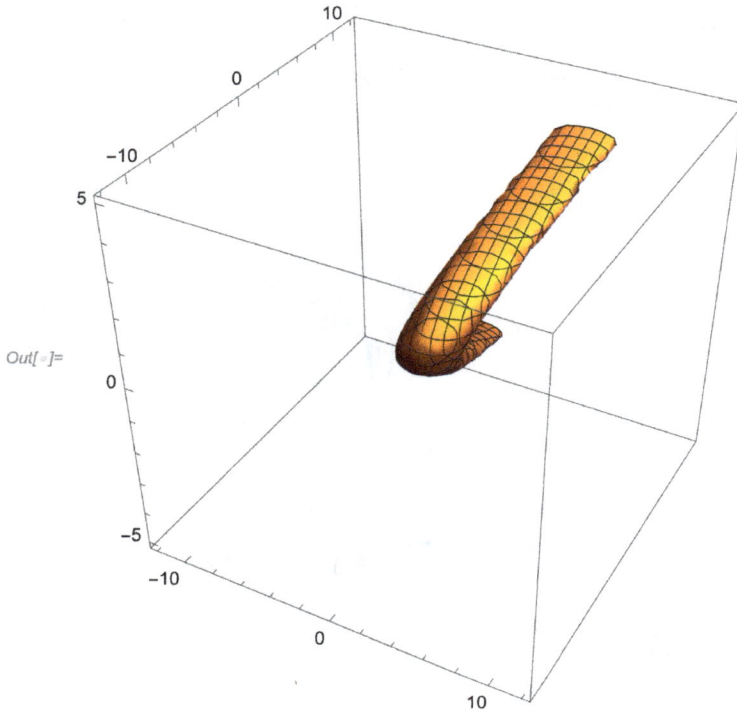

Figure 4.86

In this example we have a family of circles with origins at points (t, t^2). In the next dynamic illustration, we see a circle moving on a parabola.

```
In[·]:= Manipulate[Show[ContourPlot[F[x, y, t]
         == 0, {x, -12, 12}, {y, -12, 12},
         PlotRange -> {{-6, 6}, {-6, 6}}],
         Graphics[{Red, PointSize[0.01],
         Point[{t, t^2}]}], ParametricPlot
         [{t, t^2}, {t, -5, 5}]], {t, -2.5, 2.5}]
```

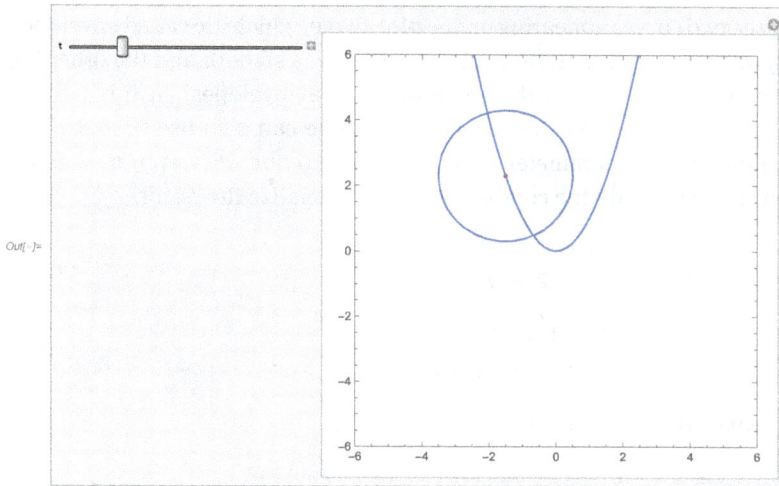

Figure 4.87

Another way to visualize is to choose the parameter *t* changing in discrete steps of 1/10 and draw the family of corresponding circles by using Table.

```
In[·]:= grenv1 = Show[Table[ContourPlot[
          F[x, y, t] == 0, {x, -12, 12},
          {y, -12, 12}], {t, -5, 5, 1/10}],
          PlotRange -> All]
```

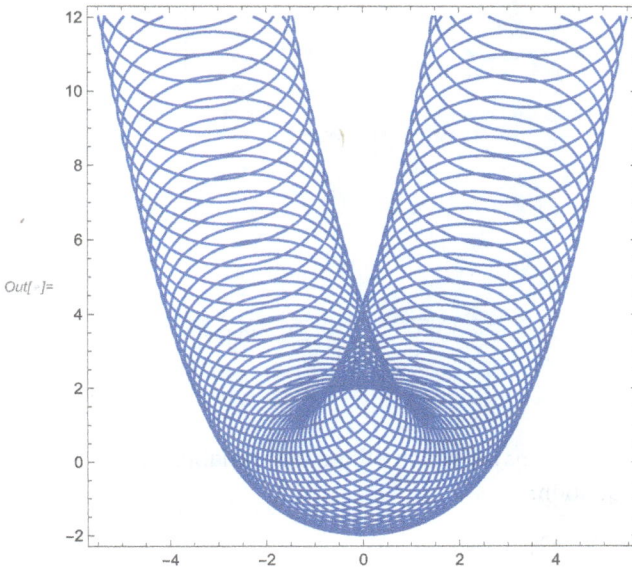

Figure 4.88

We can clearly see two curves appearing on the plot above, which are called envelopes. They are everywhere tangent to these circles. One curve is smooth and the other has complicated singularities. What is the equation for the envelopes? In this particular case it is convenient to use Groebner basis, but one can also use Resultant or Eliminate to eliminate the parameter from $F(x, y, t) = 0$ and $\partial F(x, y, t)/\partial t = 0$. Let us consider several ways to do the computations and visualize the result.

In[·]:= eq1 = F[x, y, t] == 0
Out[·]:= -4 + (-t + x)^2 + (-t^2 + y)^2 == 0

In[·]:= eq2 = D[F[x, y, t], t] == 0
Out[·]:= -2*(-t + x) - 4*t*(-t^2 + y) == 0

In[·]:= Resultant[eq1[[1]], eq2[[1]], t];

In[·]:= ContourPlot[% == 0, {x, -12, 12}, {y, -12, 12}]

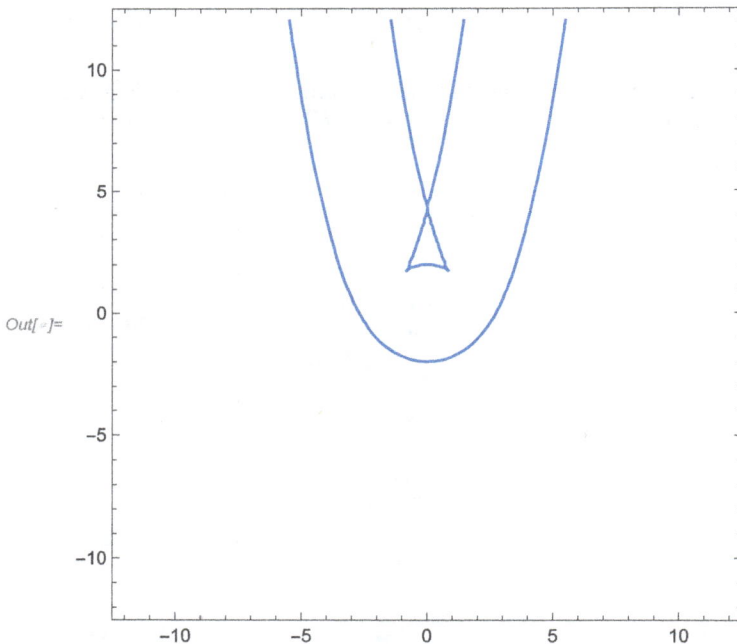

Out[·]=

Figure 4.89

We can also use the function Eliminate, which eliminates variable *t* from the two equations and returns one equation:

In[·]:= Eliminate[{eq1, eq2}, t];

We can obtain the same plot as above by typing

In[·]:= ContourPlot[Evaluate[%], {x,
    -12, 12}, {y, -12, 12}];

Another (equivalent) way to eliminate variables is to use GroebnerBasis since *F* is polynomial in its variables.

In[·]:= envel = GroebnerBasis[{F[x, y, t],
    D[F[x, y, t], t]}, {x, y}, {t}];

Let us visualize the family of circles (blue), the envelope (red), and a parabola on which the centers of the circles lie (green).

In[·]:= grenv2 = ContourPlot[envel == 0, {
    x, -12, 12}, {y, -12, 12},
    ColorFunction -> Function[{x, y}, Red]]

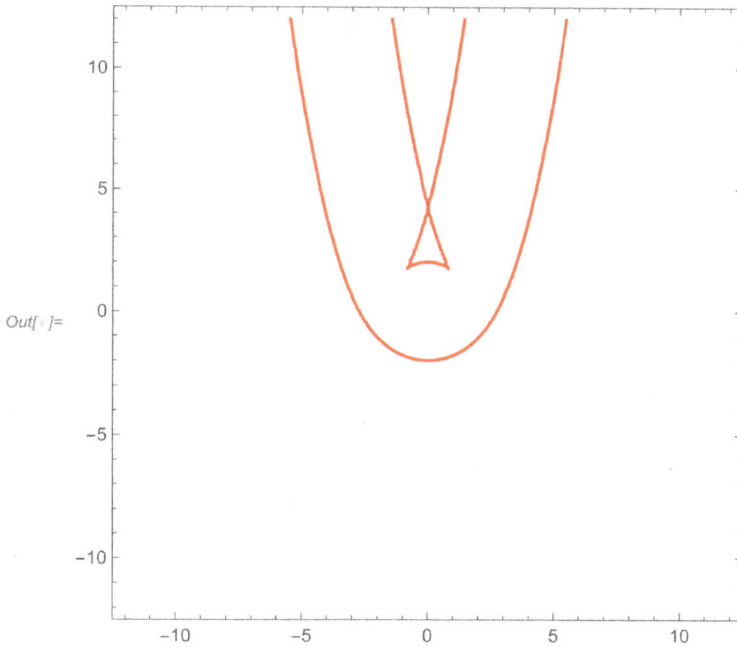

Figure 4.90

In[·]:= grenv3 = ParametricPlot[{t, t^2},
    {t, -3.5, 3.5}, PlotStyle -> Green];

In[·]:= Show[grenv1, grenv2, grenv3]

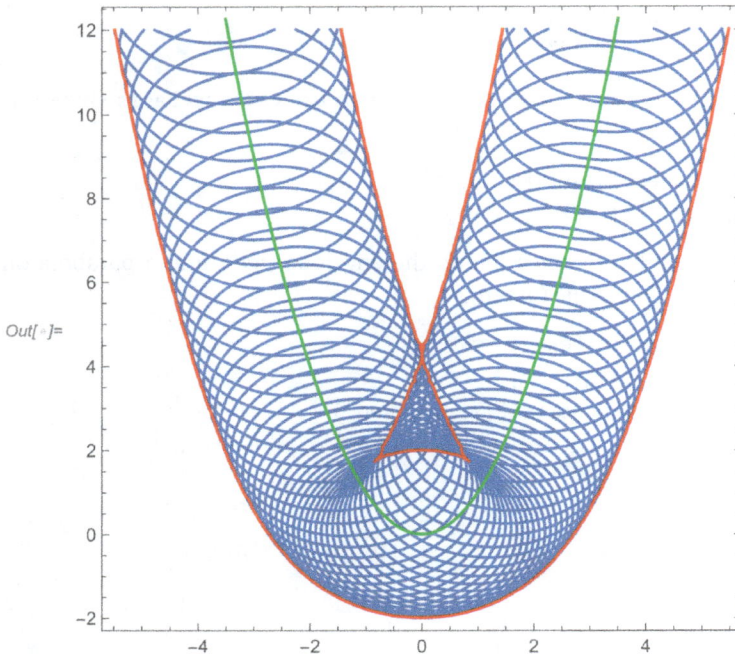

Out[ ]=

Figure 4.91

4.3.12 Curves in $\mathbb{R}^3$

Curves in $\mathbb{R}^3$ can be given by parametric equations $x = x(t)$, $y = y(t)$, $z = z(t)$ or implicitly by two equations $F_1 = F_1(x, y, z) = 0$ and $F_2 = F_2(x, y, z) = 0$. They can also be given by one equation $F_1^2 + F_2^2 = 0$. Let us take as an example the helix. It is included among the built-in SpaceCurve entities (the list of built-in space curves contains only 8 entries).

In[·]:= Entity["SpaceCurve"] // EntityList;

In[·]:= Length[%]
Out[·]:= 8

In[·]:= Entity["SpaceCurve", "Helix"]
          ["ParametricEquations"]

Out[ ]= Function$\left[\{r, c\}, \text{Function}\left[t, \{r \text{ Cos}[t], r \text{ Sin}[t], c\,t\}\right]\right]$

Figure 4.92

We shall consider the (elliptical) helix given by

In[·]:= elhelix[t_] := {2 Cos[t], 3 Sin[t], t};

In[·]:= `elhelixplot = ParametricPlot3D[`
`    elhelix[t], {t, -4 Pi, 4 Pi},`
`    ColorFunction -> Function[{x, y, z},`
`    RGBColor[z, 0, 0]], Axes -> False]`

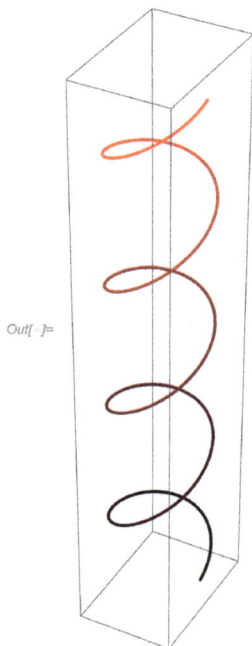

Out[·]=

Figure 4.93

The built-in function `FrenetSerretSystem` gives the generalized curvatures and Frenet–Serret basis for the parametric curve in $\mathbb{R}^n$. Recall that in two dimensions for a circle it returns the curvature, tangent, and normal vectors:

In[·]:= `FrenetSerretSystem[{Cos[t], Sin[t]}, t]`
Out[·]:= `{{1}, {{-Sin[t], Cos[t]},`
`       {-Cos[t], -Sin[t]}}}`

For a helix in $\mathbb{R}^3$, the same function returns the curvature, torsion, and the associated basis, i. e., the tangent, normal, and binormal vectors. Below we give the output for $t = 0$:

In[·]:= `Simplify[FrenetSerretSystem[elhelix[t],`
`       t], Assumptions -> Element[t, Reals]]`
`       /. t -> 0`
Out[·]:= `{{1/5, 3/20}, {{0, 3/Sqrt[10],`
`       1/Sqrt[10]}, {-1, 0, 0}, {0,`
`       -(1/Sqrt[10]), 3/Sqrt[10]}}}`

At every point of a (regular) curve, we have a tangent line. The set of all tangent lines forms an object which is called a tangent bundle. In the example below it is actually a surface. To visualize the tangent bundle, we use the function InfiniteLine:

```
In[·]:= tgspace = Graphics3D[{Blue,
        Table[InfiniteLine[elhelix[t],
        Derivative[1][elhelix][t]],
        {t, -4*Pi, 4*Pi, 0.1}]}]
```

Out[·]=

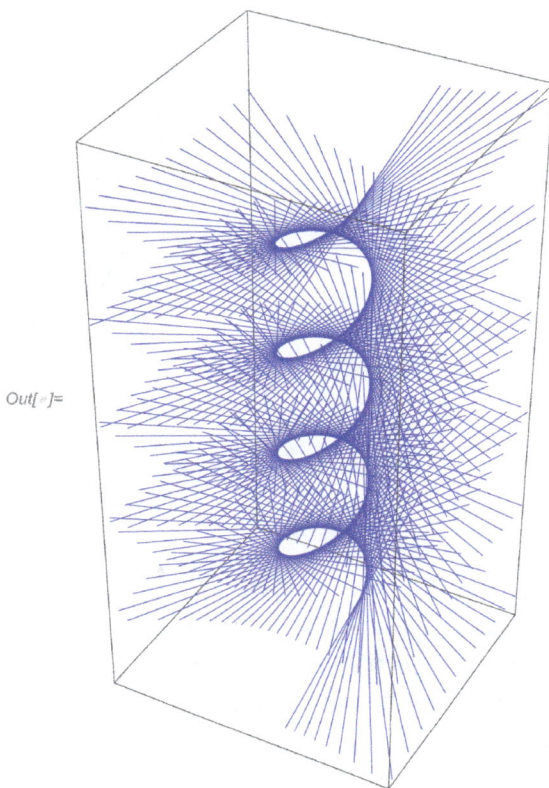

Figure 4.94

There are infinitely many planes which contain a given tangent line to a curve. Let us explain the notion of the osculating plane (see also [26]). The osculating plane to a curve y at a point P is the limit (if it exists) of the planes containing the tangent to y at P and a point Q as Q tends to P.

```
In[·]:= osculating[t_, r_] := First[z /.
        Solve[Det[{{x, y, z} - r[t],
        Derivative[1][r][t],
        Derivative[2][r][t]}] == 0, z]]
```

```
In[·]:= Manipulate[Show[Graphics3D[{
        Green, PointSize[0.02], Point[elhelix[t]],
        Line[{elhelix[t] - 10*(Derivative[1]
        [elhelix][t]/Norm[Derivative[1][elhelix]
        [t]]), elhelix[t] + 10*(Derivative[1]
        [elhelix][t]/Norm[Derivative[1]
        [elhelix][t]])}]}], elhelixplot,
        Plot3D[osculating[t, elhelix],
        {x, -10, 10}, {y, -10, 10},
        PlotRange -> {{-5, 5}, {-5, 5},
        {-16, 16}}], {t, -4*Pi, 4*Pi}]
```

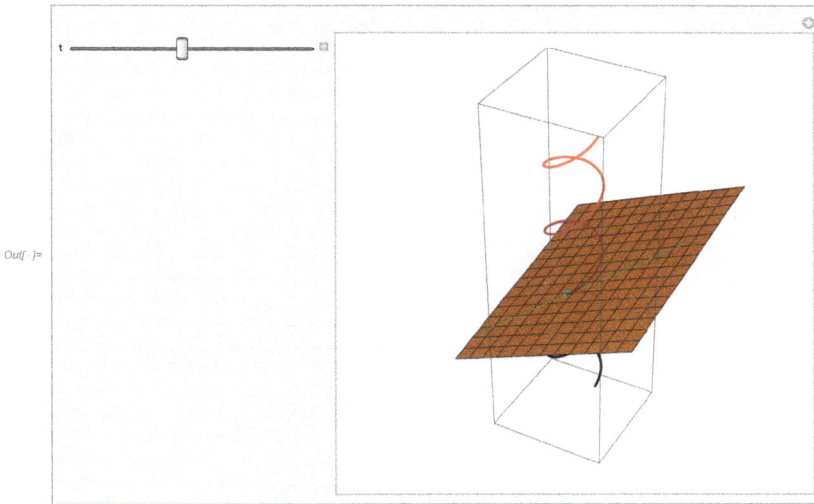

Figure 4.95

One can also think of an osculating plane as the plane which contains a tangent and a normal at a given point. Let us first define a unit tangent vector

```
In[·]:= tgv[r_][t_] := Derivative[1][r][t]/
        Sqrt[Derivative[1][r][t] . Derivative[1][r][t]]
```

and a unit normal vector

```
In[·]:= nrv[r_][t_] := Derivative[1][tgv[r]][t]
        /Sqrt[Derivative[1][tgv[r]][t] .
        Derivative[1][tgv[r]][t]]
```

The binormal vector is defined as their cross product:

```
In[·]:= biv[r_][t_] := Cross[tgv[r][t],
        nrv[r][t]]
```

The torsion, a measure of whether the curve lies in a plane, is essentially defined using the binormal vector. We can visualize all three vectors together:

```
In[·]:= Manipulate[Show[Graphics3D[{Black,
        Arrowheads[Small], Arrow[{elhelix[t],
        elhelix[t] + tgv[elhelix][t]}], Brown,
        Arrow[{elhelix[t], elhelix[t] +
        nrv[elhelix][t]}], Blue, Arrow[{
        elhelix[t], elhelix[t] + biv[
        elhelix][t]}]}], elhelixplot],
        {t, -Pi, Pi}]
```

Out[·]=

Figure 4.96

The curvature, torsion, and three vectors satisfy the so-called Frenet–Serret formulas. As we have already shown, they all can be found using the built-in function FrenetSerretSystem.

The following code will produce the same picture of the osculating plane as above:

```
In[·]:= osculatingPlane[t_, r_] :=
        ParametricPlot3D[Evaluate[r[t]
        + u*tgv[r][t] + v*nrv[r][t]],
        {u, -5, 5}, {v, -5, 5}]
```

```
In[·]:= Manipulate[Show[Graphics3D[{
        Green, PointSize[0.02], Point[
        elhelix[t]], Line[{elhelix[t] - 10*(
        Derivative[1][elhelix][t]/Norm[
        Derivative[1][elhelix][t]]),
        elhelix[t] + 10*(Derivative[1]
        [elhelix][t]/Norm[Derivative[1]
        [elhelix][t]])}]}], elhelixplot,
        osculatingPlane[t, elhelix],
        PlotRange -> {{-5, 5}, {-5, 5},
        {-16, 16}}], {t, 0, 4*Pi}]
```

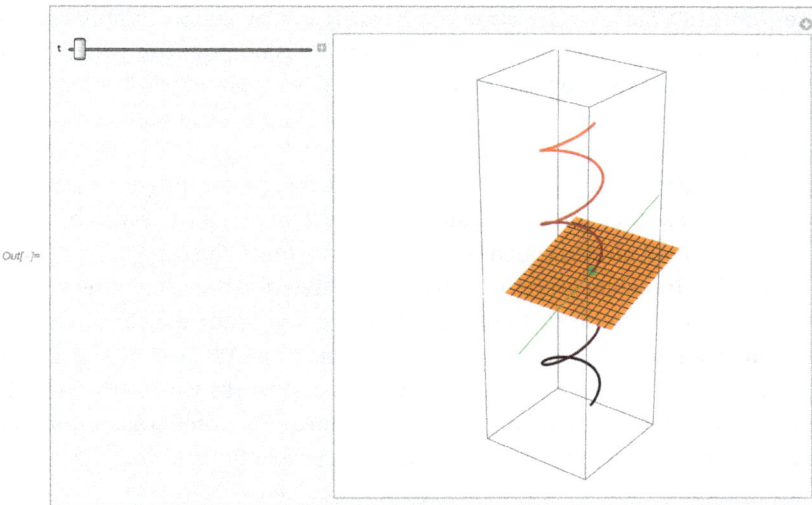

Figure 4.97

As for plane curves, we can define the notions of an osculating circle and sphere for a space curve at a given point on the curve. The osculating circle, whose curvature is the same as of the curve at a given point, lies in the osculating plane, it touches the curve and the radius is the inverse of the curvature. It has a three-point contact with the curve. One way to draw the osculating circle is to start with an ordinary circle at the origin and then translate and rotate it in the correct position in the osculating plane. Let us first define the curvature

In[·]:= curv[r_][t_] := Sqrt[Derivative[1]
        [tgv[r]][t] . Derivative[1][tgv[r]]
        [t]/Derivative[1][r][t] .
        Derivative[1][r][t]]

To collect the coefficients at *x*, *y*, *z* in the following expression which appears in the definition of the osculating plane and find the normal vector to this plane, we define the function

In[·]:= opv[t_] := Block[{x, y, z},
        Coefficient[Det[{{x, y, z} -
        elhelix[t], Derivative[1][elhelix]
        [t], Derivative[2][elhelix][t]}],
        {x, y, z}]]

In[·]:= Simplify[opv[t]]
Out[·]:= {3 Sin[t], -2 Cos[t], 6}

Let us explain the second part of the code below. First, we take a circle in the plane $z = 0$ of radius equal to the inverse of the curvature and plot it by using the function ParamtericPlot3D. Since the function GeometricTransformation is applicable to geometric objects corresponding to primitives, and there is no primitive which gives a circle in three dimensions, we want to extract the lines with which the function ParametricPlot3D draws a circle. In our case the function Cases is applied and we search for lines at all levels, as is indicated by the option Infinity. Next, we want to rotate and translate these lines, which will have an effect of performing the same operation on the circle that they form. We want the unit vector along the *z*-axis become the vector opv[t] defined above, which is the normal vector to the osculating plane. Moreover, we need to translate everything to the correct part of the osculating plane, that is, to the point corresponding to the center of the osculating circle. This is done using the function TranslationTransform. Both functions TranslationTransform and RotationTransform return TransfomationFunction which works with both numerical and symbolic vectors and essentially represents a linear fractional transformation given by certain matrices. Therefore, we can take a dot product to multiply the matrices to find the composite transformation.

In[·]:= RotationTransform[{{0, 0, 1}, opv[t]}]
        /. t -> 1 // N

Out[·]= TransformationFunction$\left[\begin{array}{ccc|c} 0.923342 & 0.0328142 & 0.382573 & 0. \\ 0.0328142 & 0.985953 & -0.163765 & 0. \\ -0.382573 & 0.163765 & 0.909296 & 0. \\ \hline 0. & 0. & 0. & 1. \end{array}\right]$

Figure 4.98

In[·]:= `TranslationTransform[elhelix[t] +`
     `1/curv[elhelix][t] nrv[elhelix][t]]`
     `/. t -> 1 // N`

Out[ ]= `TransformationFunction`$\left[\left(\begin{array}{ccc|c} 1. & 0. & 0. & -0.522576 \\ 0. & 1. & 0. & 0.651808 \\ 0. & 0. & 1. & 1.33726 \\ \hline 0. & 0. & 0. & 1. \end{array}\right)\right]$

Figure 4.99

In[·]:= `% . %%`

Out[ ]= `TransformationFunction`$\left[\left(\begin{array}{ccc|c} 0.923342 & 0.0328142 & 0.382573 & -0.522576 \\ 0.0328142 & 0.985953 & -0.163765 & 0.651808 \\ -0.382573 & 0.163765 & 0.909296 & 1.33726 \\ \hline 0. & 0. & 0. & 1. \end{array}\right)\right]$

Figure 4.100

Finally, we draw the transformed lines using the function `Graphics3D` and combine with the helix, the osculating plane, and a point in the first part of the code.

In[·]:=
```
Manipulate[Show[Graphics3D[{
    Green, PointSize[0.02], Point[
    elhelix[t]]}], elhelixplot,
    Plot3D[osculating[t, elhelix],
    {x, -20, 20}, {y, -20, 20}, Mesh
    -> False, ColorFunction ->
    Function[{x, y, z}, Directive[
    Opacity[0.2], Blue]]], Graphics3D[
    GeometricTransformation[Cases[
    ParametricPlot3D[(1/curv[elhelix]
    [t])*{Cos[u], Sin[u], 0},
    {u, 0, 2*Pi}], _Line, Infinity]
    [[1]], {TranslationTransform[
    elhelix[t] + (1/curv[elhelix][t])*
    nrv[elhelix][t]] . RotationTransform[
    {{0, 0, 1}, opv[t]}]}]}],
    PlotRange -> {{-20, 20},
    {-20, 20}, {-16, 16}}],
    {t, -4*Pi, 4*Pi}]
```

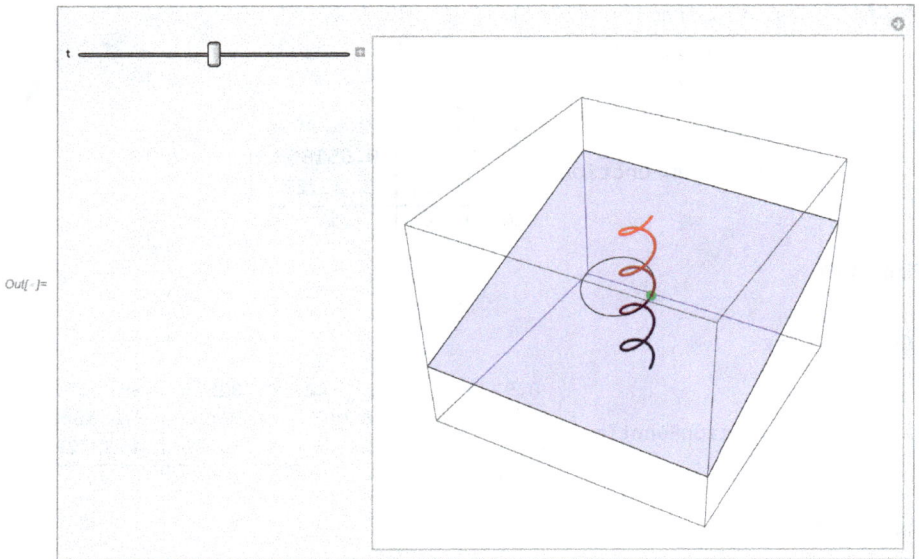

Figure 4.101

The osculating sphere for a space curve at a point on a curve is the sphere which has a four-point contact with the curve at this point. It depends on both the curvature and torsion of the space curve. Let us demonstrate this notion for a helix.

```
In[·]:= curelhelix = FrenetSerretSystem[
           elhelix[t], t][[1]][[1]];
```

```
In[·]:= torelhelix = FrenetSerretSystem[
           elhelix[t], t][[1]][[2]];
```

```
In[·]:= sigmaeh[t_] := 1/torelhelix
```

```
In[·]:= rhoeh[t_] := 1/curelhelix
```

```
In[·]:= Manipulate[Show[Graphics3D[{
           Green, PointSize[0.02], Point[
           elhelix[t1]], Yellow, Opacity[0.1],
           Evaluate[Simplify[Sphere[elhelix[t]
           + rhoeh[t]*nrv[elhelix][t] +
           sigmaeh[t]*Derivative[1][rhoeh][t]*
           biv[elhelix][t], Sqrt[rhoeh[t]^2 +
           sigmaeh[t]^2*Derivative[1][rhoeh]
           [t]^2]] /. t -> t1]]}], elhelixplot,
           PlotRange -> {{-20, 20},
           {-20, 20}, {-30, 30}}], {{t1,
           -4*Pi, "t"}, -4*Pi, 4*Pi}]
```

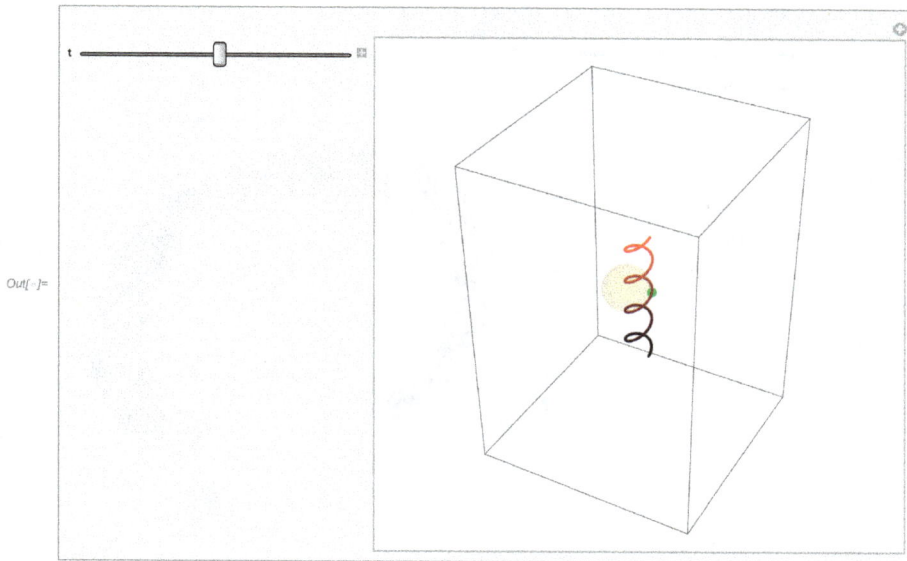

Out[·]=

Figure 4.102

4.4 Surfaces

In this chapter we shall consider smooth surfaces given by parametric equations. Most of definitions and further examples can be found in [1]. In fact, we shall consider what is known as a patch, a mapping from an open set of $\mathbb{R}^2$ (for instance, a rectangle) to $\mathbb{R}^3$. We shall consider regular patches without singularities where the Jacobian vanishes.

A mapping $g : U \to \mathbb{R}^n$, where $U \subset \mathbb{R}^2$ and $n \geq 3$, can be thought of as describing a piece of a surface in the Euclidean space $\mathbb{R}^n$. For $n = 3$, it can be displayed by using Mathematica®'s function `ParametricPlot3D`. For example, let $g(u, v) = ((\cos(u) + 4)\sin(v), \sin(u), (\cos(u) + 4)\cos(v))$ and let $U = [-a, a] \times [-b, b]$ be a rectangle in $\mathbb{R}^2$ (where we take both a and b between 0 and π). The surface we obtain is a piece of a torus.

```
In[·]:= Manipulate[ParametricPlot3D[{(4 + Cos[u])
        Sin[v], Sin[v], (4 + Cos[u]) Cos[v]},
        {u, -a, a}, {v, -b, b}, Axes -> False],
        {{a, Pi, "a"}, -Pi, Pi}, {{b, Pi,
        "b"}, -Pi, Pi}]
```

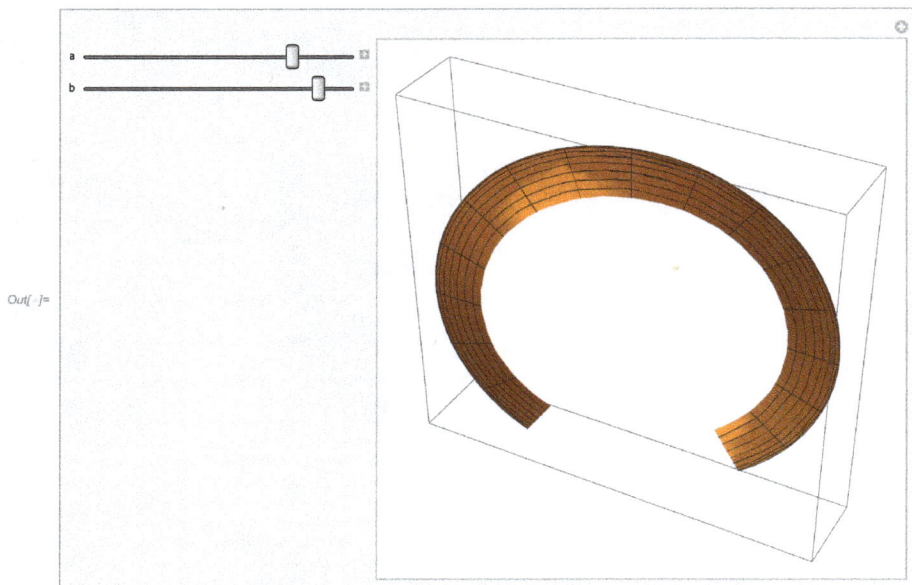

Out[·]=

Figure 4.103

4.4.1 Drawing surfaces using curves

In most cases it is sufficient to draw surfaces using ParametricPlot3D (when the surface is given parametrically by $x = x(u, v)$, $y = y(u, v)$, $z = z(u, v)$) and ContourPlot3D (when it is given implicitly by an equation $F(x, y, z) = 0$). Another way to draw surfaces in Mathematica® is to use curves and functions like Tube or RotationTransform. Let us consider a couple of examples.

```
In[·]:= elhelix[t]
Out[·]:= {2 Cos[t], 3 Sin[t], t}
```

```
In[·]:= plotparhelix = ParametricPlot3D[
        elhelix[t], {t, -4 Pi, 4 Pi},
        ColorFunction -> Function[{x, y, z},
        RGBColor[1, 0, 0]], Axes -> False]
```

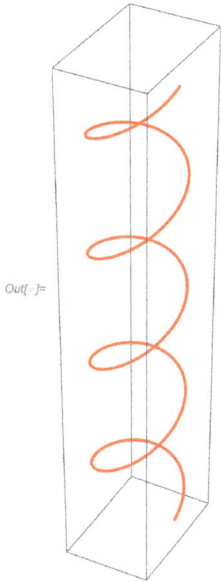

Figure 4.104

```
In[·]:= Show[Normal[plotparhelix] /.
        Line[x__] :> Tube[x, 1], PlotRange
        -> All]
```

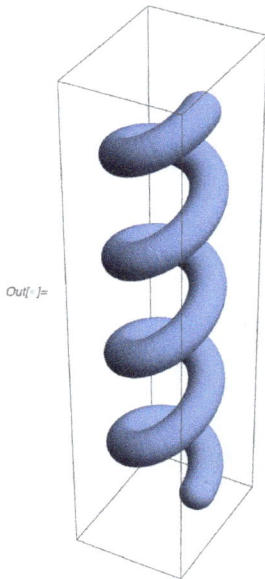

Figure 4.105

In[·]:= ```Graphics3D[{Tube[{{-1, -1, -1},
      {1, 1, 1}}, 0.2]}]```

Out[·]=

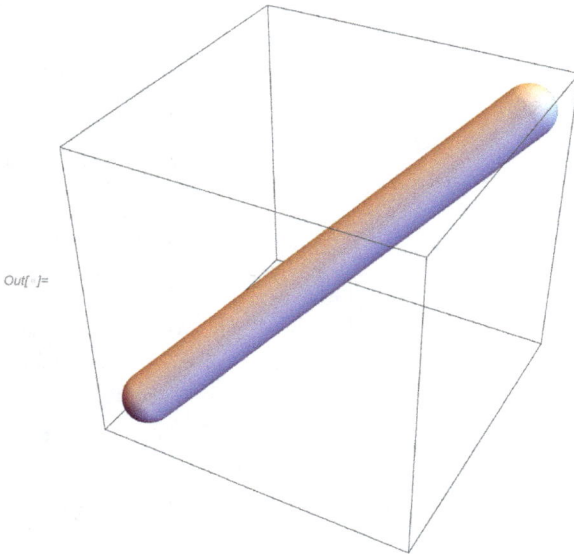

Figure 4.106

In[·]:= ```Graphics3D[{CapForm[None], Tube
      [{{-1, -1, -1}, {1, 1, 1}}, 0.2]}]```

Out[·]=

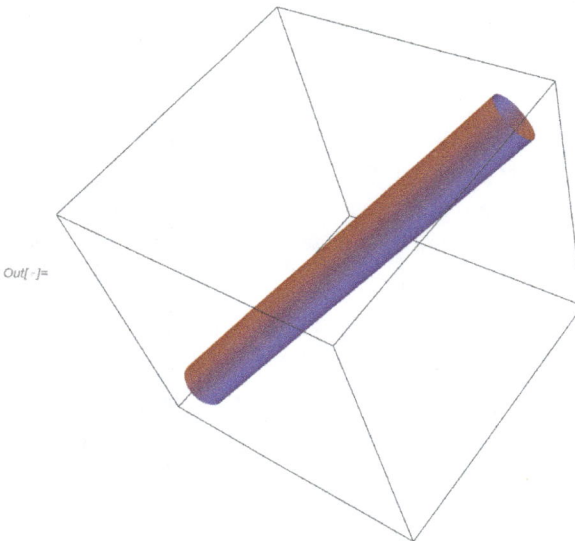

Figure 4.107

In[.]:= `Graphics3D[{CapForm["Butt"], Tube`
`    [{{-1, -1, -1}, {1, 1, 1}}, 0.2]}]`

Out[ ]=

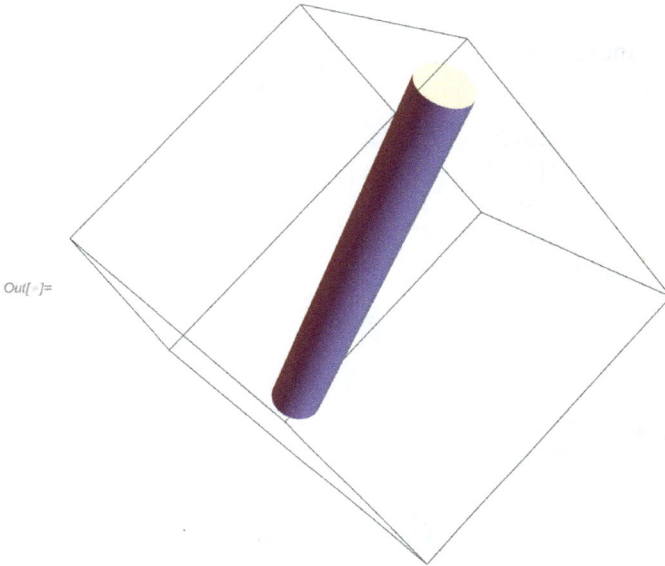

Figure 4.108

Another example is as follows:

In[.]:= `circle3d[t_] := {0, Sin[t], Cos[t] + 4}`

In[.]:= `ParametricPlot3D[circle3d[t],`
`    {t, -Pi, Pi}, ColorFunction ->`
`    (Red &), AxesLabel -> {"x", "y", "z"},`
`    ImageSize -> Small, Ticks -> False]`

Out[ ]=

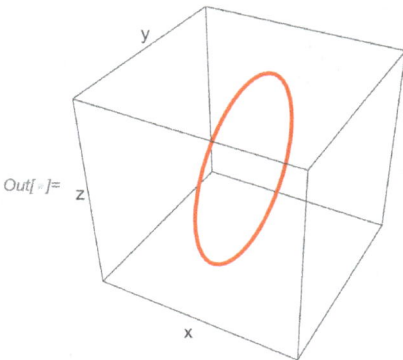

Figure 4.109

In[·]:= Tor[u_, v_] := RotationTransform
        [v, {0, 1, 0}][circle3d[u]]

In[·]:= Tor[u, v]
Out[·]:= {(4 + Cos[u]) Sin[v], Sin[u],
        (4 + Cos[u]) Cos[v]}

In[·]:= torusplot = ParametricPlot3D
        [Tor[u, v], {u, -Pi, Pi}, {v, -Pi, Pi},
        Axes -> False, Mesh -> None,
        ColorFunction -> (Directive[
        Opacity[0.4], Green] &)]

Out[·]=

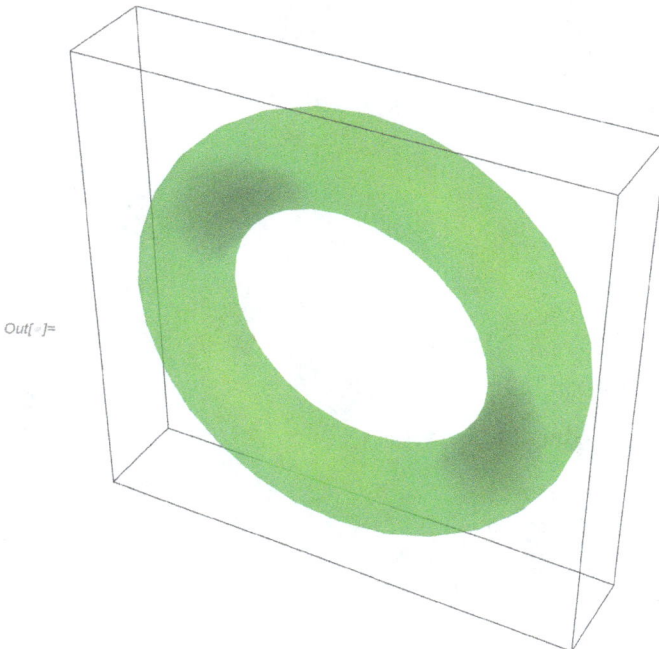

Figure 4.110

One can easily get the graphics from Entity:

In[·]:= torent = Entity["Surface", "Torus"]
        ["Graphics3D"]

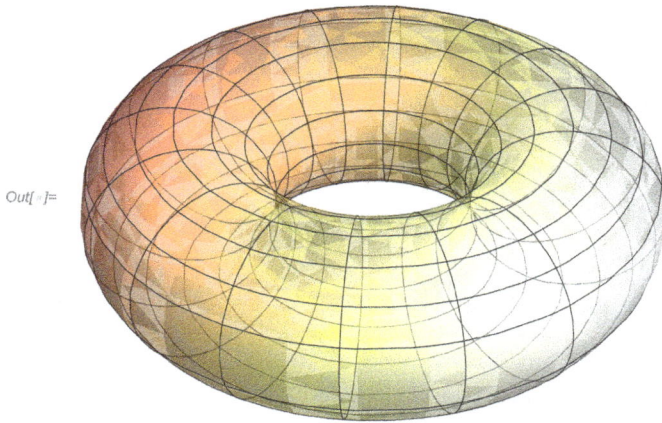

Figure 4.111

In[·]:= Show[torent, Boxed -> True]

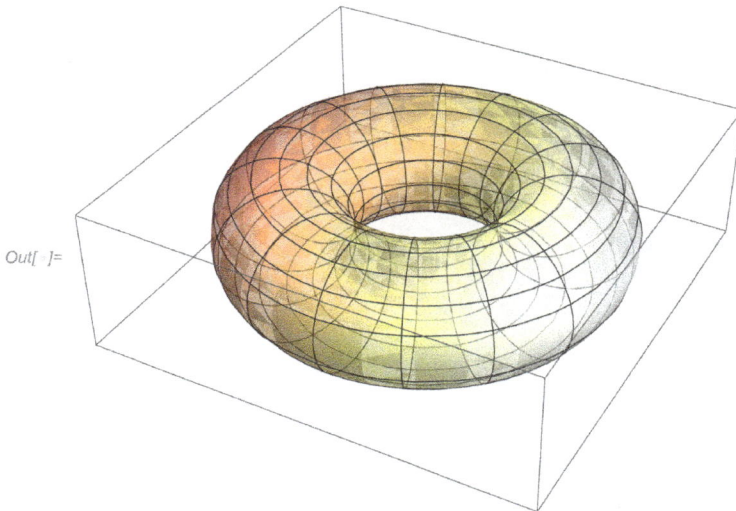

Figure 4.112

We can have different surfaces of revolution obtained from a given curve.

In[·]:= Plot[t^4 - t^2, {t, 0, 1}]

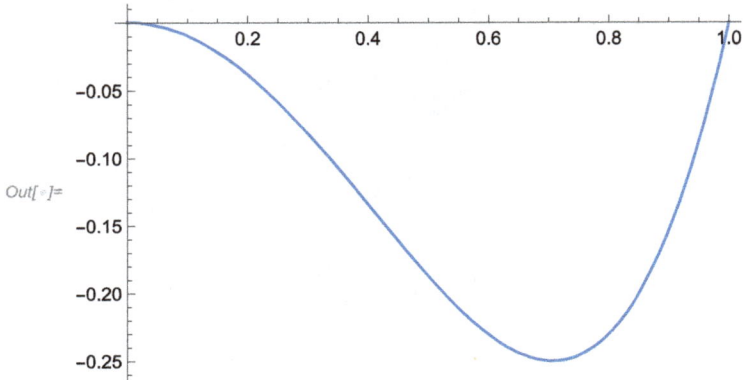

Out[ ]=

Figure 4.113

In[·]:= RevolutionPlot3D[t^4 - t^2, {t, 0, 1}]

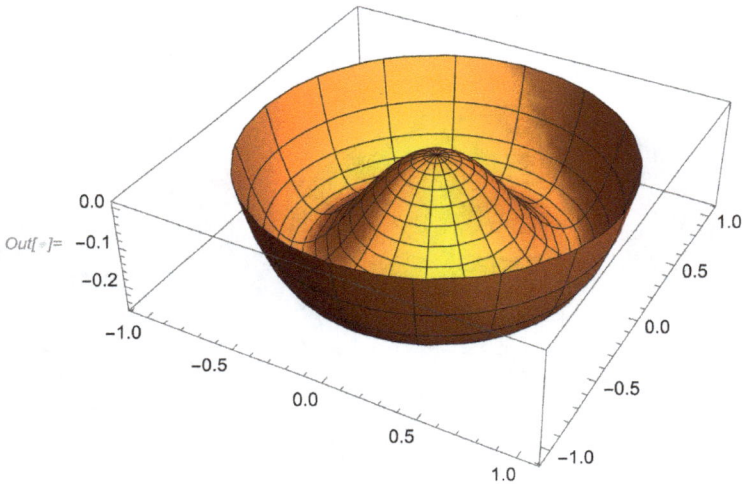

Out[ ]=

Figure 4.114

We can add cosine and sine functions at corresponding places to get a rotation around particular coordinate axes, for instance,

In[·]:= func[z_] := z

In[·]:= ParametricPlot3D[{func[z] Cos[t],
        func[z] Sin[t], z}, {z, -1, 1},
        {t, 0, 2 Pi}]

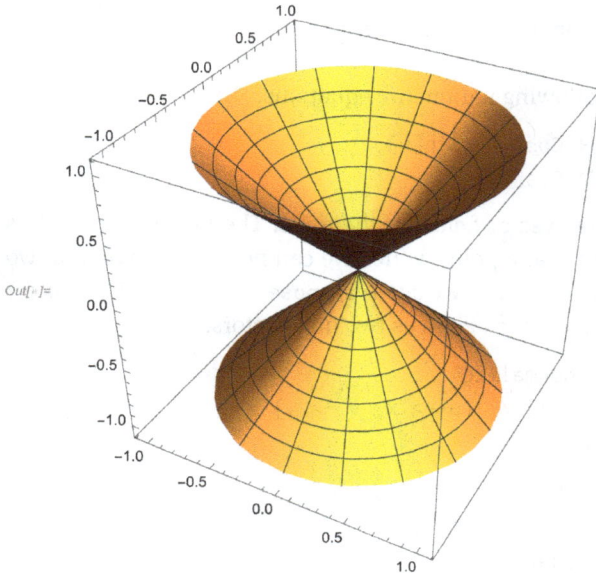

Figure 4.115

```
In[·]:= ParametricPlot3D[{func[z] Cos[t],
        z, func[z] Sin[t]}, {z, -10, 10},
        {t, 0, 2 Pi}]
```

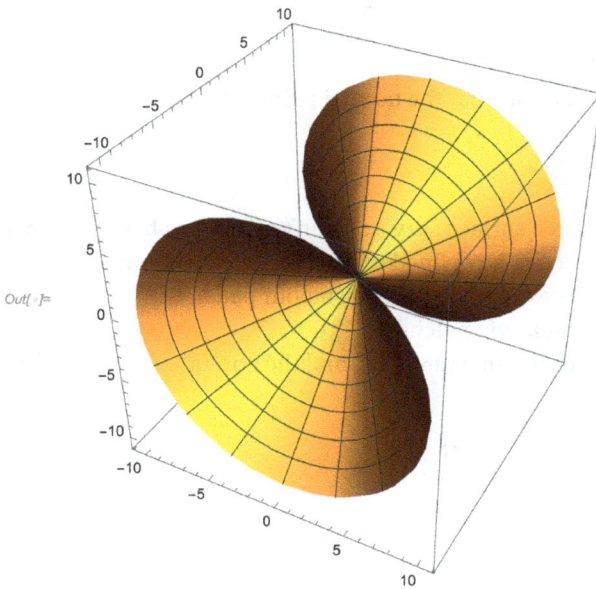

Figure 4.116

4.4.2 A tangent space and a normal to a smooth surface

Consider a torus given by the following parametric equations:

```
In[·]:= torp[u_, v_] := {(4 + Cos[u]) Sin[v],
        Sin[u], (4 + Cos[u]) Cos[v]}
```

Let us define the following three vector fields on the torus. The Jacobi matrix gives two tangent vectors to the surface at a point, which we can normalize to obtain two unit tangent vectors. In the following code we use Transpose for the Jacobi matrix for convenience and the function Normalize gives us two unit vectors:

```
In[·]:= tgp[f_][{x_, y_}] := Normalize
        /@ (D[f[u, v], {{u, v}}] /. {u -> x, v -> y}
        // Transpose)
```

```
In[·]:= tgp[torp][{1, 2}] // N
Out[·]:= {{-0.765147, 0.540302, 0.350175},
          {-0.416147, 0., -0.909297}}
```

```
In[·]:= Norm[#] & /@ %
Out[·]:= {1., 1.}
```

These vectors define a tangent plane to the smooth surface at a point. To define a normal which is perpendicular to the tangent plane at a point, we use a cross product:

```
In[·]:= nrp[f_][{x_, y_}] := -Normalize[Cross
        @@ (Transpose[D[f[u, v], {{u, v}}]])]
        /. {u -> x, v -> y}
```

The minus sign is chosen in such a way that the normal points outside.

```
In[·]:= nrp[torp][{Pi, -Pi}]
Out[·]:= {0, 0, 1}
```

The normal measures how much the surface curves. A surface may not have a normal vector field defined everywhere. It can be defined on a compact closed surface only if it is orientable. But locally it can be defined over any smooth patch. So here we shall consider surfaces only locally (that is, via patches).

Let us use Arrow and Tube to draw tangent vectors and the normal vector at a given point.

```
In[·]:= tgpVects[f_][{u_, v_}] := Arrow[
        Tube[{f[u, v], f[u, v] + #}, .02]]
        & /@ tgp[f][{u, v}]
```

In[·]:= `normpVect[f_][{u_, v_}] := Arrow[`
        `Tube[{f[u, v], f[u, v] + nrp[f]`
        `[{u, v}]}, .02]]`

The arrow goes from a point on the surface to the same point plus the normal vector.

We shall visualize the torus (green), tangent vectors (blue), and a normal (red), and using a 2D slider on the left we can choose a point on the torus where u and v vary from $-\pi$ to π:

In[·]:= `ttrp = ParametricPlot3D[torp[u, v],`
        `{u, -Pi, Pi}, {v, -Pi, Pi}, Axes ->`
        `False, Mesh -> None, ColorFunction ->`
        `(Directive[Opacity[0.4], Green] &)];`

In[·]:= `Manipulate[Show[ttrp, Graphics3D[{`
        `Thick, Red, Arrowheads[Small],`
        `normpVect[torp][p], Blue, tgpVects`
        `[torp][p]}], PlotRange -> {{-6, 6},`
        `{-6, 6}, {-6, 6}}], {{p, {0, 0}},`
        `{-Pi, -Pi}, {Pi, Pi}, Slider2D}]`

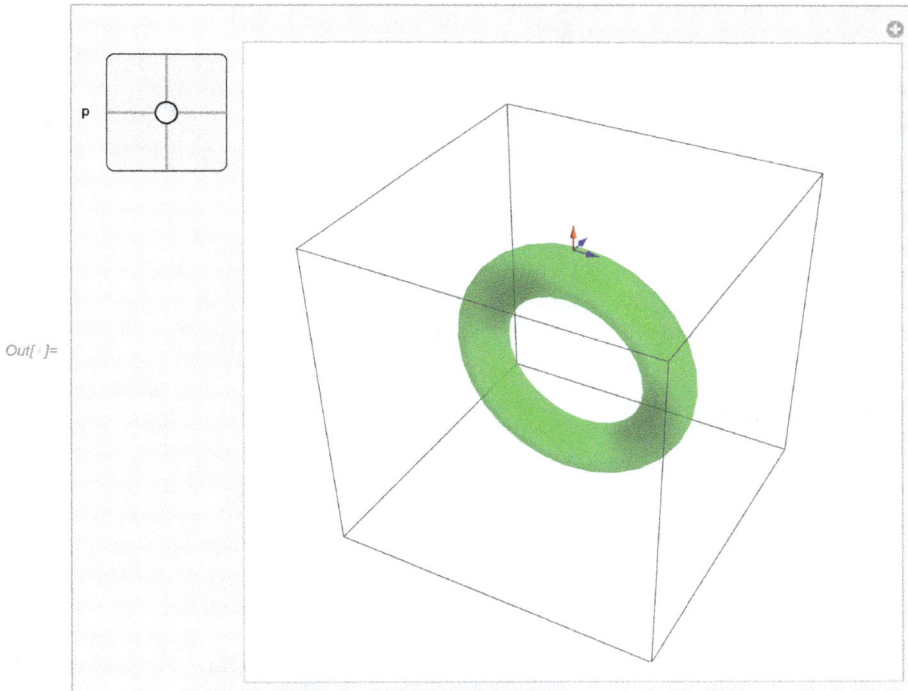

Out[·]=

Figure 4.117

4.4.3 Metrics on a surface: The first and second fundamental forms

Using metrics we can measure distance on a surface. Let $f : D \to \mathbb{R}^3$ be a coordinate patch for a surface $S \subset \mathbb{R}^3$ and D be an open disk in $\mathbb{R}^2$.

4.4.3.1 The first fundamental form

In order to define the metric form $Edu^2 + 2Fdudv + Gdv^2$ on D, we shall introduce functions E1, F1, and G1 as follows. The coefficients E, F, and G are called the components of the first fundamental form.

```
In[·]:= E1[f_][u_, v_] := Simplify[
         Derivative[1, 0][f][u, v] .
         Derivative[1, 0][f][u, v]]
```

```
In[·]:= F1[f_][u_, v_] := Simplify[
         Derivative[1, 0][f][u, v] .
         Derivative[0, 1][f][u, v]]
```

```
In[·]:= G1[f_][u_, v_] := Simplify[
         Derivative[0, 1][f][u, v] .
         Derivative[0, 1][f][u, v]]
```

We can also define the area element (or the infinitesimal area)

```
In[·]:= infArea[f_][u_, v_] := Sqrt[
         E1[f][u, v]*G1[f][u, v] -
         F1[f][u, v]^2]
```

The metric form is defined by

```
In[·]:= metricForm[f_][u_, v_] :=
         E1[f][u, v]*du^2 + 2*F1[f][u, v]*
         du*dv + G1[f][u, v]*dv^2
```

and for the torus it is

```
In[·]:= metricForm[torp][u, v]
Out[·]:= du^2 + (4 + Cos[u])^2*dv^2
```

The infinitesimal area is

```
In[·]:= infArea[torp][u, v] // PowerExpand
Out[·]:= 4 + Cos[u]
```

For a surface which is included in the Entity dataset, we can obtain all this information from Mathematica®.

```
In[·]:= Entity["Surface", "Torus"]["AreaElement"]
```

Out[ ]= Function$\left[\{a, c\}, \text{Function}\left[\{u, v\}, a \text{ Abs}\left[c + a \text{ Cos}\left[v\right]\right]\right]\right]$

Figure 4.118

In[ ]:= Entity["Surface", "Torus"]["FirstFundamentalForm"]

Out[ ]= Function$\left[\{a, c\}, \text{Function}\left[\{u, v\}, \left\{\left(c + a \text{ Cos}\left[v\right]\right)^2, 0, a^2\right\}\right]\right]$

Figure 4.119

In[ ]:= Entity["Surface", "Torus"]["SecondFundamentalForm"]

Out[ ]= Function$\left[\{a, c\}, \text{Function}\left[\{u, v\}, \left\{-\left(\left(c + a \text{ Cos}\left[v\right]\right) \text{Cos}\left[v\right]\right), 0, -a\right\}\right]\right]$

Figure 4.120

For a sphere, we have

In[ ]:= sphp[r_][u_, v_] := r {Cos[u] Cos[v],
        Cos[v] Sin[u], Sin[v]}

In[ ]:= ParametricPlot3D[sphp[1][u, v],
        {u, 0, 2*Pi}, {v, -Pi/2, Pi/2}]

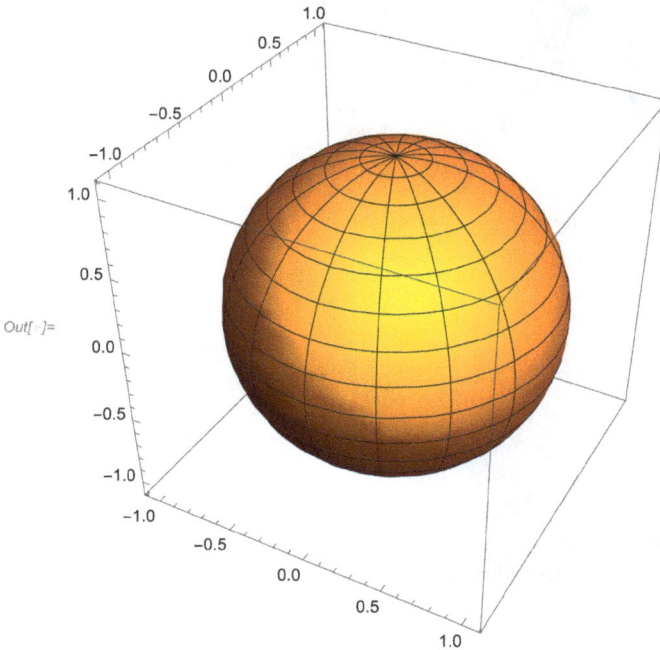

Figure 4.121

```
In[·]:= metricForm[sphp[r]][u, v]
Out[·]:= r^2*Cos[v]^2*du^2 + r^2*dv^2
```

Integrating over the whole surface, we find a well-known formula for the surface area of the sphere:

```
In[·]:= Assuming[r > 0, Simplify[Integrate[
        infArea[sphp[r]][u, v], {u, 0, 2*Pi},
        {v, -Pi/2, Pi/2}]]]
Out[·]:= 4*Pi*r^2
```

For a monkey saddle,

```
In[·]:= surf[s_, t_] := {s, t, t^2 - s^2}
```

```
In[·]:= ParametricPlot3D[surf[s, t],
        {s, -1, 1}, {t, -1, 1}]
```

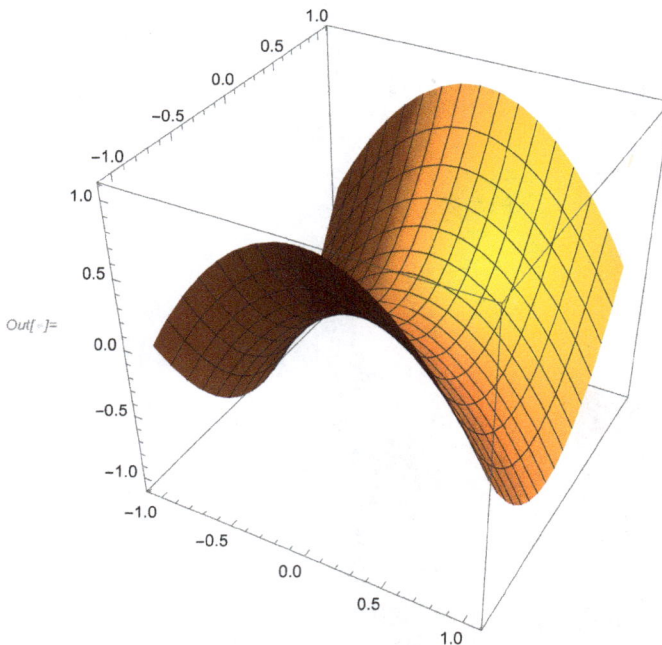

Figure 4.122

```
In[·]:= metricForm[surf][u, v]
Out[·]:= (1 + 4*u^2)*du^2 - 8*u*v*du*dv
        + (1 + 4*v^2)*dv^2
```

The infinitesimal area is

```
In[·]:= infArea[surf][u, v] // Simplify
Out[·]:= Sqrt[1 + 4*u^2 + 4*v^2]
```

For a torus with parameters defined by

```
In[·]:= ttp[a_, b_][s_, t_] := {Cos[t] (b +
        a Cos[s]), Sin[t] (b + a Cos[s]),
        a Sin[s]}
```

```
In[·]:= Manipulate[ParametricPlot3D[
        ttp[a, b][s, t], {s, 0, 2 Pi},
        {t, 0, 2 Pi}], {{a, 1, "a"}, 0.1, 3},
        {{b, 1, "b"}, 0.1, 3}]
```

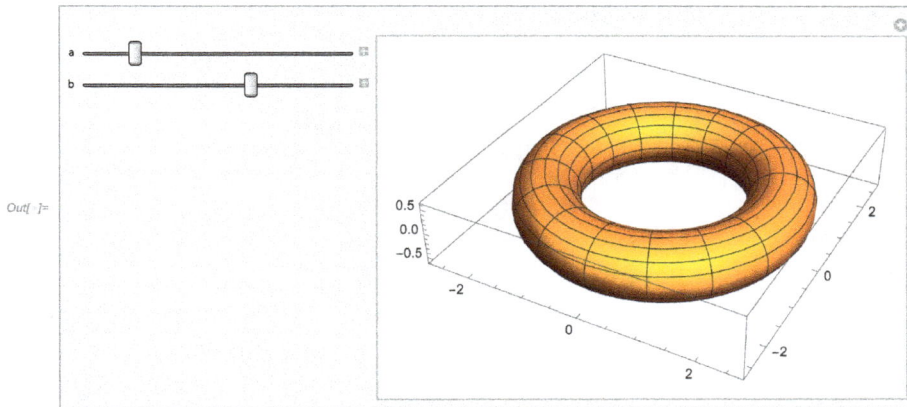

Figure 4.123

```
In[·]:= Simplify[infArea[ttp[a, b]][u, v],
        {a > 0, b > a, , 0 < u < 2 Pi}]
Out[·]:= a (b + a Cos[u])
```

```
In[·]:= Integrate[infArea[ttp[a, b]][u, v],
        {u, 0, 2 Pi}, {v, 0, 2 Pi}, Assumptions
        -> {a > 0, b > 0, b > a}]
Out[·]:= 4*a*b*Pi^2
```

4.4.3.2 The second fundamental form

Usually the coefficients of the second fundamental form are denoted by e, f, g, but in the code below we shall use e2, f2, and g2. They are defined by rather complicated formulas which involve the second-order derivatives and the coefficients of the first fundamental form, which we defined above by E1, F1, and G1.

```
In[·]:= e2[f_][u_, v_] := Simplify[Det[
        {Derivative[2, 0][f][u, v],
        Derivative[1, 0][f][u, v],
        Derivative[0, 1][f][u, v]}]/
        Sqrt[E1[f][u, v]*G1[f][u, v]
        - F1[f][u, v]^2]]
```

```
In[·]:= f2[f_][u_, v_] := Simplify[Det[
        {Derivative[1, 1][f][u, v],
        Derivative[1, 0][f][u, v],
        Derivative[0, 1][f][u, v]}]/
        Sqrt[E1[f][u, v]*G1[f][u, v]
        - F1[f][u, v]^2]]
```

```
In[·]:= g2[f_][u_, v_] := Simplify[Det[
        {Derivative[0, 2][f][u, v],
        Derivative[1, 0][f][u, v],
        Derivative[0, 1][f][u, v]}]/
        Sqrt[E1[f][u, v]*G1[f][u, v]
        - F1[f][u, v]^2]]
```

```
In[·]:= g2[ttp[1, 1]][0, 4]
Out[·]:= 2
```

For a smooth surface, one can define the concept of normal curvature which carries information about how much the surface is curving in each direction. To compute its value at a point, we need to know a normal vector at a point and some tangent vector in the direction of which we want to compute normal curvature. We take a normal section in the direction of the chosen tangent vector and obtain a curve. Normal curvature in this direction is defined as the curvature of this curve. If we consider all possible directions at a given point, there must be maximum and minimum values of normal curvature among them. These are called the principal curvatures k_1, k_2. Using the principal curvatures on a surface, one can define the Gaussian curvature $K = k_1 k_2$ and the mean curvature $H = (k_1 + k_2)/2$. It is a remarkable fact that the Gaussian curvature does not depend on the normal vector field, it is intrinsic and depends only on the metric of the surface. It can be proved by finding a formula for K in terms of the coefficients of the first fundamental form only. If one deforms the surface locally isometrically (preserving distances) then the Gaussian curvature will remain the same, even though the normal vector may actually change. Usually in books, the formula for K is given in terms of the coefficients of the first and second fundamental forms, but as mentioned above there is a formula that needs only the coefficients of the first fundamental form. We define the Gaussian curvature by

```
In[·]:= gcurvature[x_][u_, v_] :=
        Simplify[(Det[{Derivative[2, 0]
        [x][u, v], Derivative[1, 0][x]
        [u, v], Derivative[0, 1][x][u, v]}]
        *Det[{Derivative[0, 2][x][u, v],
        Derivative[1, 0][x][u, v],
        Derivative[0, 1][x][u, v]}] -
        Det[{Derivative[1, 1][x][u, v],
        Derivative[1, 0][x][u, v],
        Derivative[0, 1][x][u, v]}]^2)/
        (Derivative[1, 0][x][u, v] .
        Derivative[1, 0][x][u, v]*
        Derivative[0, 1][x][u, v] .
        Derivative[0, 1][x][u, v] -
        Derivative[1, 0][x][u, v] .
        Derivative[0, 1][x][u, v]^2)^2]
```

```
In[·]:= gcurvature[sphp[1]][0, 0]
Out[·]:= 1
```

For the mean curvature, we use the following formula, which is quite long but makes calculations faster:

```
In[·]:= mcurvature[x_][u_, v_] :=
        Simplify[(Derivative[0, 1][x][u, v]
        . Derivative[0, 1][x][u, v]*
        Det[{Derivative[2, 0][x][u, v],
        Derivative[1, 0][x][u, v],
        Derivative[0, 1][x][u, v]}] -
        2*Derivative[1, 0][x][u, v] .
        Derivative[0, 1][x][u, v]*
        Det[{Derivative[1, 1][x][u, v],
        Derivative[1, 0][x][u, v],
        Derivative[0, 1][x][u, v]}] +
        Derivative[1, 0][x][u, v] .
        Derivative[1, 0][x][u, v]*Det[
        {Derivative[0, 2][x][u, v],
        Derivative[1, 0][x][u, v],
        Derivative[0, 1][x][u, v]}])/
        (2*(Derivative[1, 0][x][u, v] .
        Derivative[1, 0][x][u, v]*
        Derivative[0, 1][x][u, v] .
        Derivative[0, 1][x][u, v] -
        Derivative[1, 0][x][u, v] .
        Derivative[0, 1][x][u, v]^2)^(3/2))]
```

In[·]:= mcurvature[sphp[1]][0, 0] // FullSimplify
Out[·]:= -1

Let us define a torus with three parameters

In[·]:= torus3[a_, b_, c_][u_, v_] :=
        {(a + b Cos[v]) Cos[u], (a + b Cos[v])
        Sin[u], c Sin[v]}

In[·]:= tr3 = ParametricPlot3D[torus3
        [1, 0.6, 1][u, v], {u, 0, 2 Pi},
        {v, 0, 2 Pi}, Mesh -> None,
        ColorFunction -> Function[
        {x, y, z}, Directive[Opacity[0.2],
        Red]]]

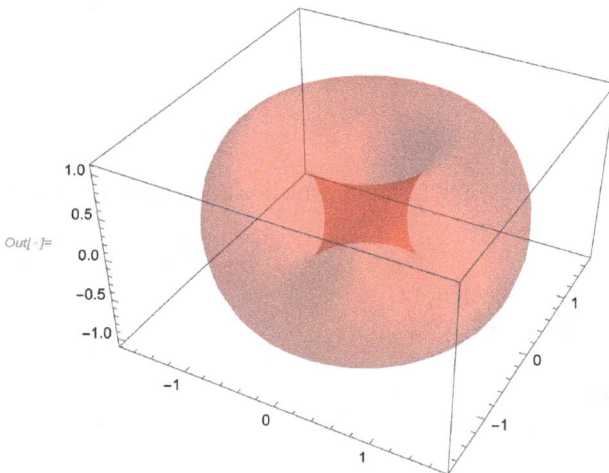

Figure 4.124

On the next dynamic illustration, we can see the Gaussian curvature at a point on a torus:

In[·]:= Manipulate[Show[Graphics3D[{
        Blue, PointSize[0.02], Point[
        torus3[1, 0.6, 1][u, v]], Style
        [Text[gcurvature[torus3[1, 0.6, 1]]
        [u, v], torus3[1, 0.6, 1][u, v] +
        {0.1, 0.1, 0.1}], {Black, Thick}]}],
        tr3], {{u, 0, "u"}, 0, 2*Pi,
        Appearance -> "Labeled"}, {{
        v, 0, "v"}, 0, 2*Pi, Appearance ->
        "Labeled"}]

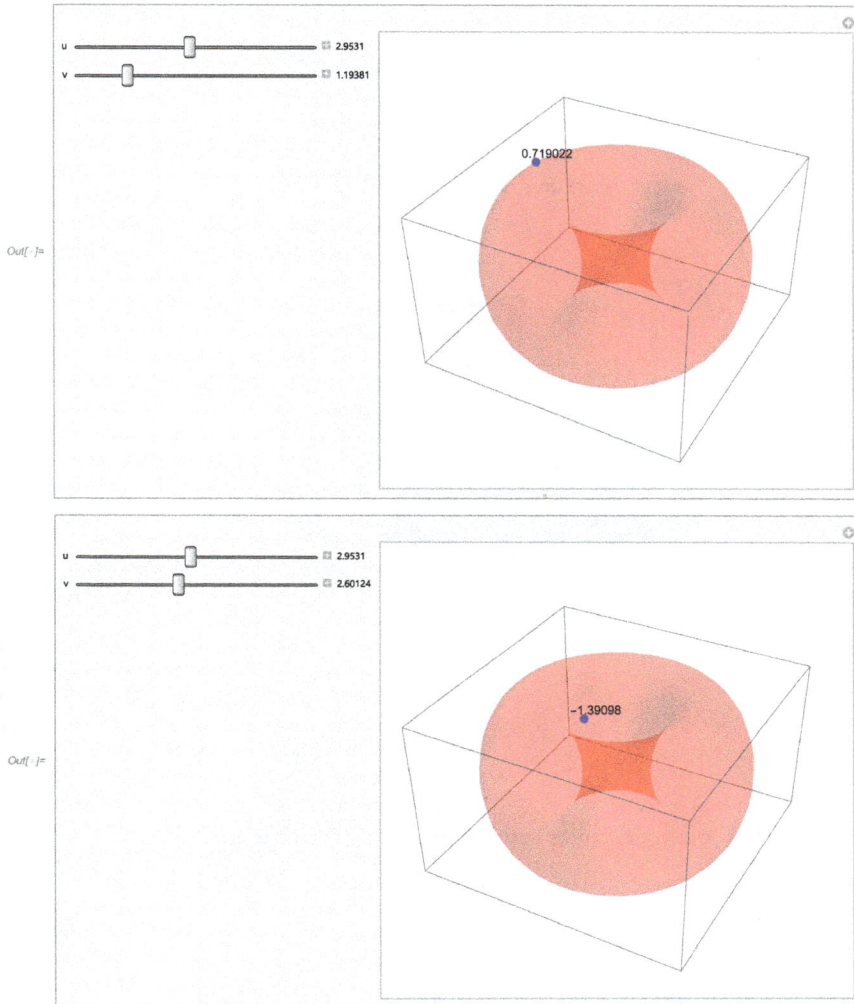

Figure 4.125

As we can see, it can be both positive and negative. Let us color the torus by curvature (positive vs negative). We shall use Hue for which the argument should be from 0 to 1.

```
In[·]:= gcolorfunction1[u_, v_] := Which[
        gcurvature[torus3[1, 0.6, 1]][u, v]
        > 0, 0.5, gcurvature[torus3[1, 0.6, 1]]
        [u, v] < 0, 1, True, 0.1];
```

In[·]:= tr1c = ParametricPlot3D[torus3
    [1, 0.6, 1][u, v], {u, 0, 2 Pi},
    {v, 0, 2 Pi}, Mesh -> None,
    ColorFunctionScaling -> False,
    ColorFunction -> Function[{x, y,
    z, u, v}, Hue[gcolorfunction1[u, v]]]]

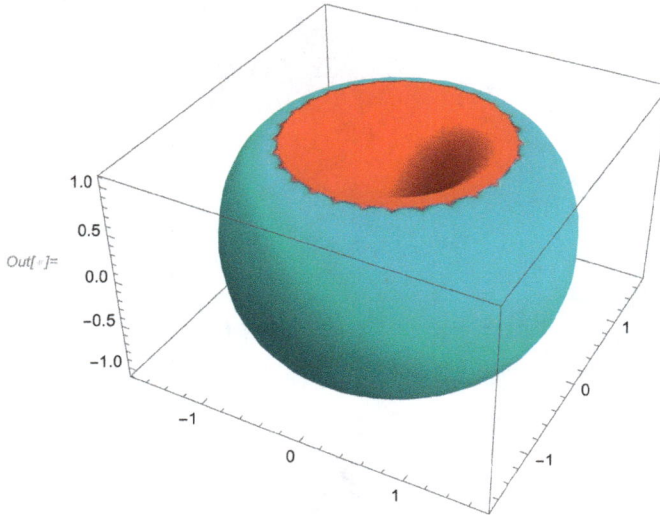

Out[·]=

Figure 4.126

The red part of the torus corresponds to negative Gaussian curvature.

In[·]:= Manipulate[Show[{Graphics3D[{
    PointSize[0.02], Point[torus3[1,
    gcurvature[torus3[1, 0.6, 1]][u, v],
    torus3[1, 0.6, 1][u, v] + {0.1,
    0.1, 0.1}], Black]}], tr1c}],
    {{u, 0, "u"}, 0, 2*Pi, Appearance
    -> "Labeled"}, {{v, 0, "v"},
    0, 2*Pi, Appearance -> "Labeled"}]

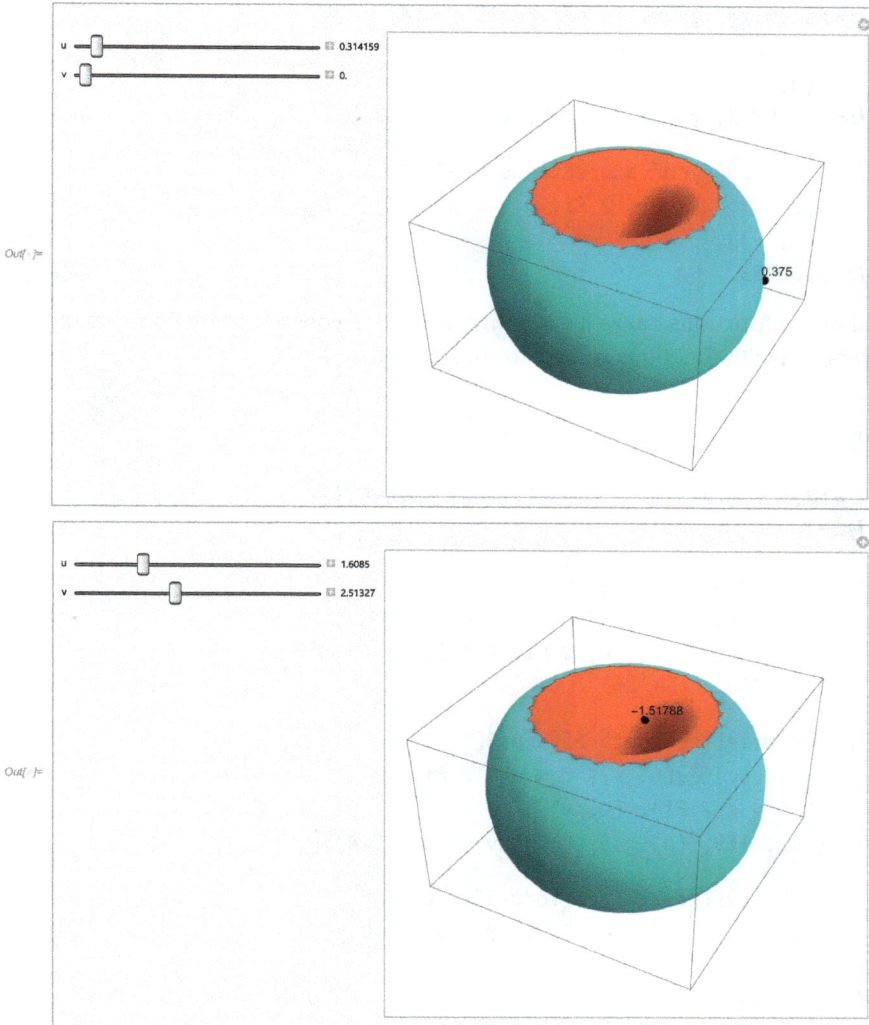

Figure 4.127

We can also color the torus by curvature in a continuous way. Let us calculate approximately the maximum and minimum values of the Gaussian curvature.

```
In[·]:= max = Max[Table[gcurvature[torus3
        [1, 0.6, 1]][u, v], {u, 0, 2 Pi, 0.1},
        {v, 0, 2 Pi, 0.1}]]
Out[·]:= 3.89382
```

We could have used the function Maximize in the following way, however, using Table above gives a good approximation and is simpler.

```
In[·]:= Maximize[{gcurvature[torus3[1, 0.6,
          1]][u, v], 0 < u < 2 Pi, 0 < v < 2 Pi},
          {u, v}]
Out[·]:= {3.89433, {u -> 5.49779, v -> 4.88098}}
```

```
In[·]:= MaxValue[{gcurvature[torus3[1, 0.6, 1]]
          [u, v], 0 < u < 2 Pi, 0 < v < 2 Pi},
          {u, v}]
Out[·]:= 3.89433
```

Note that the functions MaxValue and MinValue can work with functions having parameters and with functions subject to constraints:

```
In[·]:= MaxValue[a*x^2 + c, x]
Out[·]:= Piecewise[{{c, a <= 0}}, Infinity]
```

```
In[·]:= MaxValue[2*x*y, Element[{x, y}, Disk[]]]
Out[·]:= 1
```

```
In[·]:= min = Min[Table[gcurvature[
          torus3[1, 0.6, 1]][u, v], {u, 0,
          2 Pi, 0.1}, {v, 0, 2 Pi, 0.1}]]
Out[·]:= -12.2485
```

```
In[·]:= Minimize[{gcurvature[torus3[
          1, 0.6, 1]][u, v], 0 < u < 2 Pi,
          0 < v < 2 Pi}, {u, v}]
Out[·]:= {-12.4163, {u -> 5.49779, v -> 4.45524}}
```

```
In[·]:= MinValue[{gcurvature[torus3[1, 0.6,
          1]][u, v], 0 < u < 2 Pi, 0 < v < 2 Pi},
          {u, v}]
Out[·]:= -12.4163
```

We do not need exact numbers for min and max. They are used in the function Rescale with arguments x and {min, max} below which gives x rescaled to run from 0 to 1 over the range min to max. We rescale only one argument of RGBColor function.

```
In[·]:= gcolorfunction2[u_, v_] := Which[
          gcurvature[torus3[1, 0.6, 1]][u, v] > 0,
          RGBColor[Rescale[gcurvature[torus3[
          1, 0.6, 1]][u, v], {0, max}], 0.4, 0.4],
          gcurvature[torus3[1, 0.6, 1]][u, v] < 0,
          RGBColor[0.4, 0.4, Rescale[-gcurvature[
          torus3[1, 0.6, 1]][u, v], {0, -min}]],
          True, RGBColor[0, 0, 0]]
```

```
In[·]:= ParametricPlot3D[torus3[1, 0.6, 1]
        [u, v], {u, 0, 2 Pi}, {v, 0, 2 Pi}, Mesh
        -> None, ColorFunctionScaling ->
        False, ColorFunction -> Function[{x,
        y, z, u, v}, gcolorfunction2[u, v]]]
```

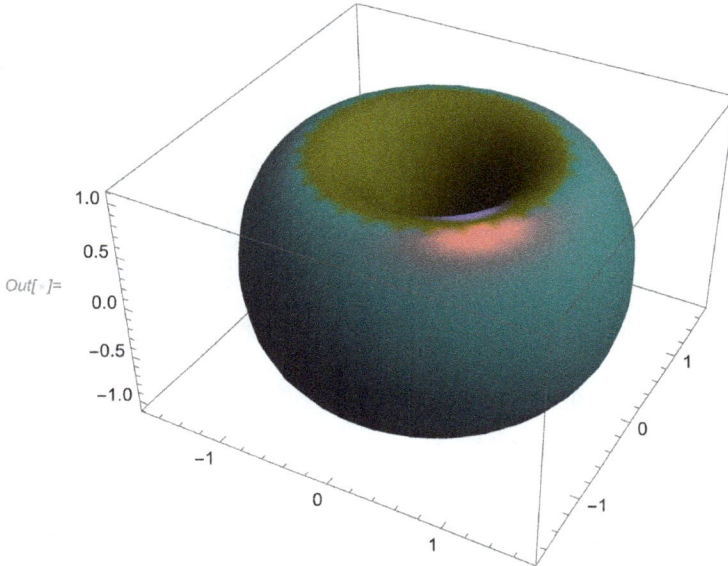

Figure 4.128

4.4.4 Curves and frames on a torus

Consider the following torus:

```
In[·]:= torusg[a_ : 1, b_ : 2, c_ : 1][u_, v_] :=
        {Cos[v] (b + a Cos[u]), Sin[v]
        (b + a Cos[u]), c Sin[u]}
```

```
In[·]:= torusg[][u, v]
Out[·]:= {(2 + Cos[u]) Cos[v], (2 + Cos[u])
        Sin[v], Sin[u]}
```

Let us take any plane curve. We will use a circle $(\cos t, \sin t)$, but any parametrized curve can be used in the same way. We can plot this curve in space or on the torus by substituting its parametrization in this way:

```
In[·]:= cv1 = ParametricPlot3D[torusg[][
        Cos[t], Sin[t]], {t, 0, 2 Pi}, ColorFunction ->
        (Red &)]
```

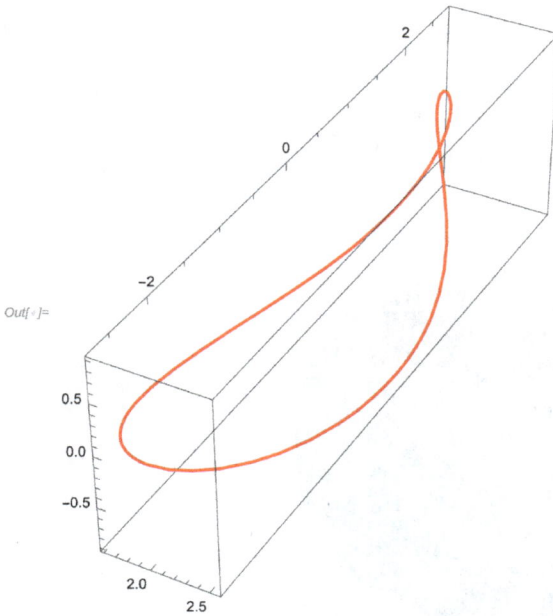

Figure 4.129

```
In[·]:= trg = ParametricPlot3D[torusg[][u, v],
        {u, 0, 2 Pi}, {v, 0, 2 Pi}, ColorFunction
        -> ({Opacity[0.4], Green} &),
        Mesh -> False];
```

```
In[·]:= Show[trg, cv1, Axes -> False]
```

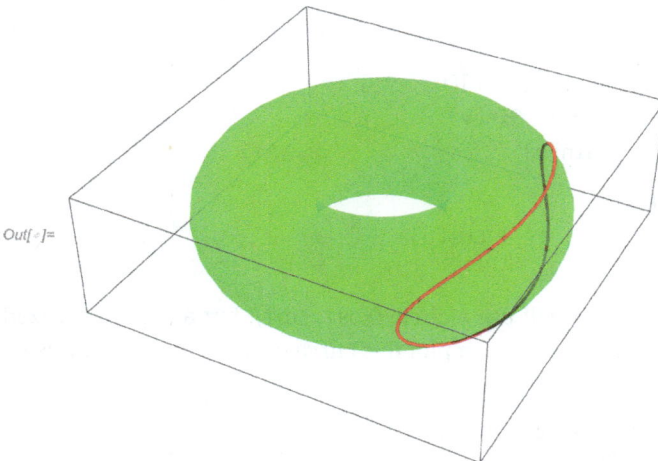

Figure 4.130

Let us define the equation of our curve:

```
In[·]:= curvet[t_] = torusg[][Cos[t], Sin[t]]
Out[·]:= {(2 + Cos[Cos[t]]) Cos[Sin[t]], (2 +
        Cos[Cos[t]]) Sin[Sin[t]], Sin[Cos[t]]}
```

We can consider the curve as an ordinary curve in space and use the built-in function FrenetSerretSystem, which returns a list consisting of the tangent vector to the curve, the normal vector which is defined using acceleration, and the binormal vector, all perpendicular to each other:

```
In[·]:= vecs[t_] = Last[FrenetSerretSystem
        [curvet[t], t]];
```

A tangent vector to the curve is also tangent to the surface. Now we can see the Frenet–Serret frame moving on the curve on the torus:

```
In[·]:= Manipulate[Show[trg, cv1, Graphics3D
        [Map[Arrow[{curvet[t], curvet[t] + #}]
        &, vecs[t]]], PlotRange -> {{-3, 3},
        {-3, 3}, {-3, 3}}], {t, 0, 2 Pi}]
```

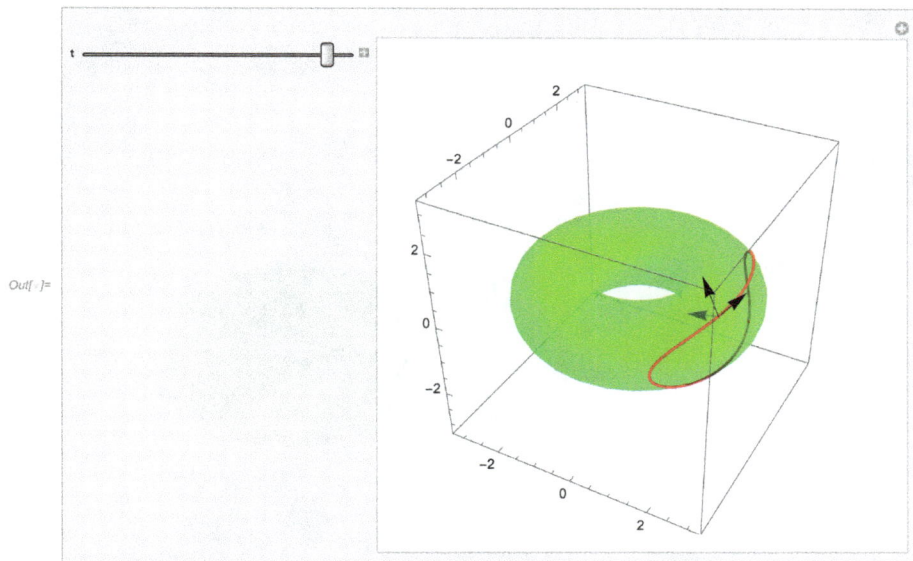

Figure 4.131

Next we consider a Darboux frame, which is defined differently. We must first compute the unit normal field to the torus:

```
In[·]:= nrg[u_, v_] = Simplify[Normalize[Apply
         [Cross, Transpose[D[torusg[][u, v],
         {{u, v}}]]]], Element[{u, v}, Reals]];
```

A Darboux frame on a curve is given by three vectors, namely t, $n \times t$, and n, where t is a unit tangent vector to the curve, and n the unit normal to the surface.

```
In[·]:= tgg[t_] := Simplify[Normalize[
         Derivative[1][curvet][t]],
         Element[t, Reals]]
```

```
In[·]:= vects1[t_] := {tgg[t], nrg[Cos[t],
         Sin[t]], Cross[nrg[Cos[t], Sin[t]],
         tgg[t]]}
```

```
In[·]:= Manipulate[Show[trg, cv1, Graphics3D
         [{Blue, Map[Arrow[{curvet[t], curvet
         [t] + #}] &, vects1[t]]}], PlotRange ->
         {{-3, 3}, {-3, 3}, {-3, 3}}],
         {t, 0, 2 Pi}]
```

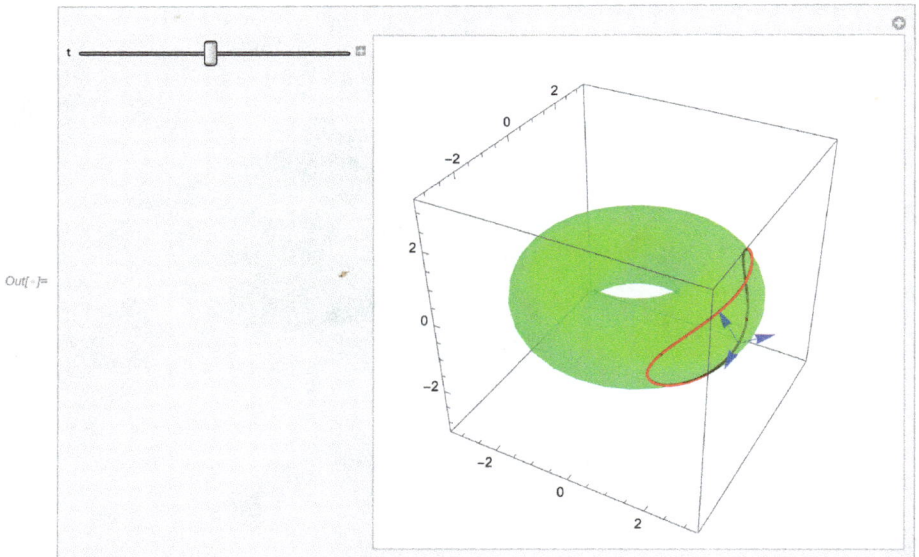

Figure 4.132

We can put both frames together using the same Manipulate:

```
In[·]:= Manipulate[Show[trg, cv1, Graphics3D
        [{If[fs, (Arrow[{curvet[t], curvet[t]
        + #1}] & ) /@ vecs[t], {}], If[
        dar, {Blue, (Arrow[{curvet[t],
        curvet[t] + #1}] & ) /@ vects1[t]
        }, {}]}], PlotRange -> {{-3, 3},
        {-3, 3}, {-3, 3}}], {t, 0, 2*Pi},
        {{fs, True, "Frenet-Serret"},
        {True, False}}, {{dar, True,
        "Darboux"}, {True, False}}]
```

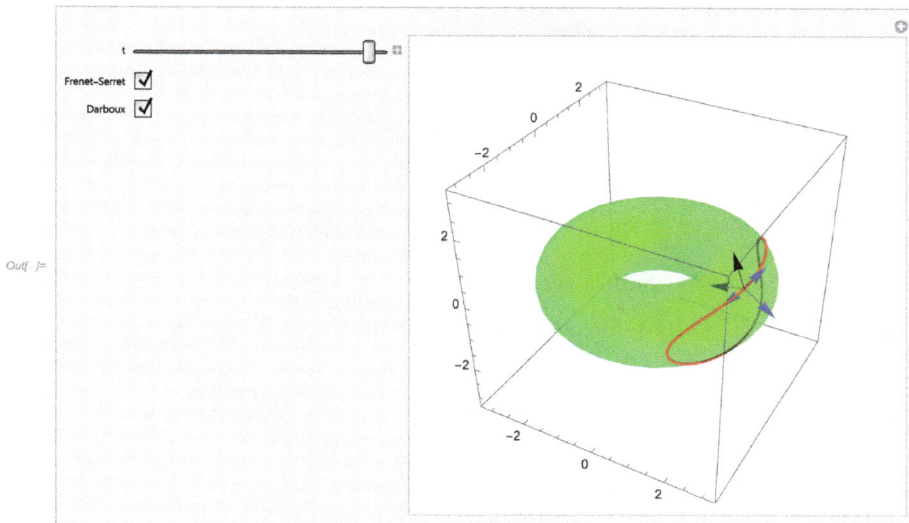

Figure 4.133

The geodesics considered below will be curves along which both frames coincide except possibly for the direction of normals. Locally a geodesic gives the shortest path
between two points on a surface.

4.4.5 Geodesics

In this section we will use code based on that in [1]. We have slightly modified it to
make it more compact. We need to compute first Christoffel symbols and then solve
certain differential equations.

In[·]:= christoffel[x_][u_, v_] :=
        Module[{A = Array[0 & , {2, 2,
        2}], E1, F1, G1, U, V}, E1 =
        D[x[U, V], U] . D[x[U, V], U]; F1 =
        D[x[U, V], U] . D[x[U, V], V]; G1
        = D[x[U, V], V] . D[x[U, V], V];
        A[[1, 1, 1]] = D[E1, V]*F1 - 2*D[F1, U]
        *F1 + D[E1, U]*G1; A[[2, 2, 2]] =
        D[G1, V]*E1 - 2*D[F1, V]*F1 +
        D[G1, U]*F1; A[[2, 1, 1]] = (-D[E1,
        V])*E1 + 2*D[F1, U]*E1 - D[E1,
        U]*F1; A[[1, 2, 2]] = (-D[G1, V])*F1
        + 2*D[F1, V]*G1 - D[G1, U]*G1;
        A[[1, 1, 2]] = D[E1, V]*G1 - D[G1, U]*F1;
        A[[2, 1, 2]] = (-D[E1, V])*F1 +
        D[G1, U]*E1; (Do[A[[i, 2, 1]] =
        A[[i, 1, 2]], {i, 1, 2}]; A/
        (E1*G1 - F1^2)/2 /. {U -> u,
        V -> v})]

To compute Christoffel symbols, we use:

In[·]:= G = Simplify[christoffel[torusg[]][0, 0]]
Out[·]:= {{{0, 0}, {0, 0}}, {{0, 0}, {0, 0}}}

Individual symbols can then be given by

In[·]:= G[[1, 1, 1]]
Out[·]:= 0

Next, we define the unit normal vector function:

In[·]:= unitnormal[x_][u_, v_] := Simplify
        [Normalize[Cross[Derivative[1, 0][x]
        [u, v], Derivative[0, 1][x][u, v]]],
        Element[_, Reals]]

In[·]:= unitnormal[torusg[]][u, v]
Out[·]:= {-Cos[u] Cos[v], -Cos[u] Sin[v], -Sin[u]}

It gives the same answer as nrg defined above:

In[·]:= nrg[u, v]
Out[·]:= {-Cos[u] Cos[v], -Cos[u] Sin[v], -Sin[u]}

The normal curvature function is defined as follows. Here x is the function parametrizing the surface and (α, β) the parameters of the plane curve that defines the curve in the plane (whose image under x is a curve on the torus).

```
In[·]:= normalcurv[x_, {a_, b_}][t_] :=
        With[{xt = x[a[#1], b[#1]] & },
        Derivative[2][xt][t] .
        unitnormal[x][a[t], b[t]]/
        Derivative[1][xt][t] .
        Derivative[1][xt][t]]
```

```
In[·]:= Simplify[normalcurv[torusg[], {
          Cos, Sin}][0.5]]
Out[·]:= 0.273283
```

The geodesic curvature is defined by

```
In[·]:= geodesiccurv[x_, {a_, b_}][t_] :=
        With[{xt = x[a[#1], b[#1]] & },
        Det[{Derivative[1][xt][t],
        Derivative[2][xt][t],
        unitnormal[x][a[t], b[t]]}]/
        (Derivative[1][xt][t] .
        Derivative[1][xt][t])^(3/2)]
```

```
In[·]:= geodesiccurv[torusg[],
          {Cos, Sin}][0.5]
Out[·]:= -0.0976597
```

Finally, here is the function that constructs several pairs of parametric functions that have to be substituted into the equation of the torus to obtain several geodesics starting from a given point. Here x is the equation of the surface, (u_0, v_0) are the coordinates of the starting point, b is the length of flow, and m the number of geodesics.

```
In[·]:= geod[x_, {u0_, v0_}, b_, m_] :=
        Module[{e, eqic, G = christoffel[x]
        [u0, v0]}, e = G[[All, 1, 1]]*
        Derivative[1][u][s]^2 + 2*G
        [[All, 1, 2]]*Derivative[1][u][s]*
        Derivative[1][v][s] + G
        [[All, 2, 2]]*Derivative[1][v][s]^2;
        eqic = {Derivative[2][u][s] +
        e[[1]] == 0, Derivative[2][v][s]
        + e[[2]] == 0, u[0] == u0, v[0]
        == v0, Derivative[1][u][0] ==
        Cos[t], Derivative[1][v][0] ==
        Sin[t]}; Flatten[Table[NDSolve[
        eqic, {u, v}, {s, 0, b}], {t,
        Pi/m, Pi, Pi/m}], 1]];
```

Let us try to find one geodesic, of the length of flow equal to 6, starting from the point with parameters $(0, 0)$. To get an actual pair of interpolating functions, we need to do this:

```
In[·]:= fg = First[{u, v} /. geod[torusg[],
       {0, 0}, 6, 1]]
```

Out[·]= {InterpolatingFunction[⊞ ╲ Domain: {{0., 6.}} Output: **scalar**], InterpolatingFunction[⊞ ── Domain: {{0., 6.}} Output: **scalar**]}

Figure 4.134

Such a pair can now be applied to values as follows:

```
In[·]:= Through[fg[0.5]]
Out[·]:= {-0.5, 0.}
```

Recall that Through with arguments p[f1, f2][x] gives p[f1[x], f2[x]]. In order to get a point on the torus, we need to use

```
In[·]:= Apply[torusg[], Through[fg[0.5]]]
Out[·]:= {2.87758, 0., -0.479426}
```

The same code can be written more efficiently as

```
In[·]:= torusg[] @@ Through[fg[0.5]]
Out[·]:= {2.87758, 0., -0.479426}
```

We can now plot our geodesic on the torus:

```
In[·]:= Show[trg, ParametricPlot3D[
       Evaluate[torusg[] @@ Through[
       First[{u, v} /. geod[torusg
       [], {0, 0}, 6, 1]][t]]],
       {t, 0, 2*Pi}], Axes -> False]
```

Out[·]=

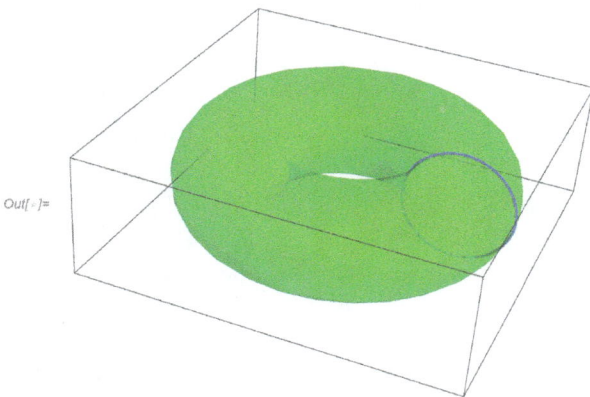

Figure 4.135

We can obtain several geodesics at once. In this case we get two:

In[·]:= `fg2 = {u, v} /. geod[torusg[],`
`        {0, 0}, 6, 2];`

Below @@@ replaces heads at the level 1.

In[·]:= `Show[trg, ParametricPlot3D[`
`        Evaluate[torusg[]] @@@ Map[`
`        Through, Through[fg2[t]]]],`
`        {t, 0, 2 Pi}], Axes -> False]`

Out[ ]=

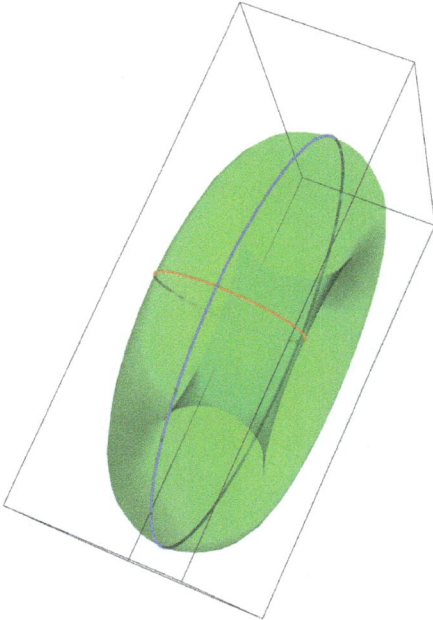

Figure 4.136

4.4.6 The tangent cone

The tangent cone is a generalization of the notion of the tangent space at a point to a smooth manifold to the case of certain spaces with singular points [23]. See also [20, 27]. The definition of a regular point on a plane curve depends on the way the curve is defined. For a curve given by $f(x, y) = 0$, we require that the gradient is nonzero. For a curve given parametrically by $x = \phi(t)$, $y = \psi(t)$, we require that the limit

$$\lim_{t \to 0} \frac{d(\phi(t), \psi(t))/dt}{\|d(\phi(t), \psi(t))/dt\|}$$

exists.

The following curves clearly have singularities:

In[·]:= `Show[{ContourPlot[Sin[x] - Cos[y] ==`
`0, {x, -6, 6}, {y, -6, 6}],`
`Graphics[{PointSize[0.01],`
`Point[{Pi/2, 0}]}]}]`

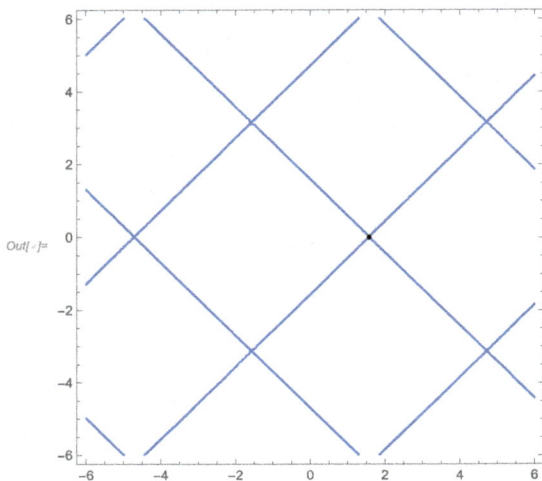

Figure 4.137

In[·]:= `ContourPlot[(x^2 + y^2 - x)^2 -`
`x^2 - y^2 == 0, {x, -1, 1},`
`{y, -1, 1}]`

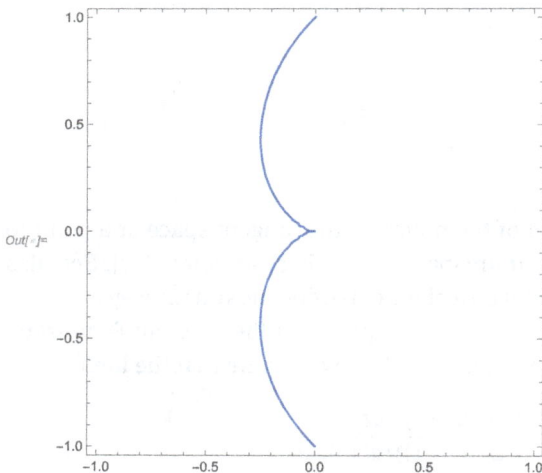

Figure 4.138

In some cases it is useful to study deformations of singularities. Let us try to find all the singular points for the entire family of curves $u^2 - 4v^2 - v^3 = a$ parametrized by the parameter a:

```
In[·]:= Reduce[{D[u^2 - 4*v^2 - v^3 - a,
        {{u, v}}] == 0, u^2 - 4*v^2 -
        v^3 == a}, {u, v}]
Out[·]:= (a == -(256/27) && u == 0 && v
        == -(8/3)) || (a == 0 && u ==
        0 && v == 0)
```

So the curve $u^2 - 4v^2 - v^3 + 256/27 = 0$ has a singular point at $(0, -8/3)$. Can we see it?

```
In[·]:= ContourPlot[u^2 - 4*v^2 - v^3 ==
        -(256/27), {u, -20, 20}, {v,
        -20, 20}, ColorFunction -> (Red & )]
```

Out[·]=

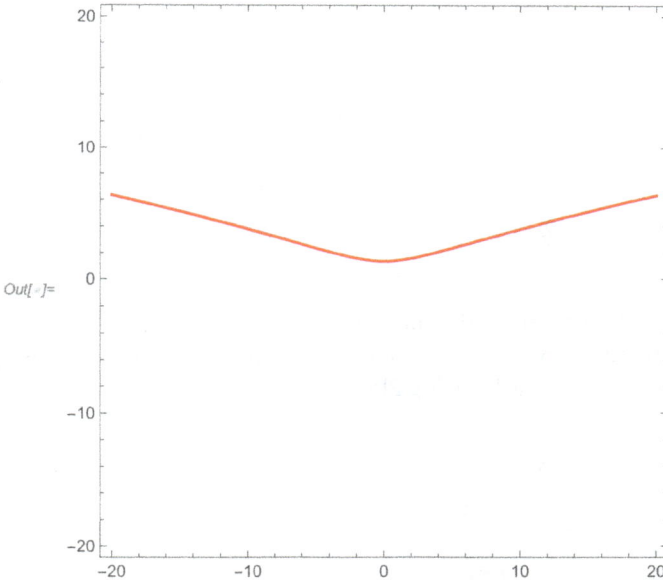

Figure 4.139

We do not see any singular point! The reason is that the curve has actually two pieces, a one-dimensional one and a zero-dimensional one. Mathematica® cannot show the lower dimensional "singular" part. (The same is true of ContourPlot3D). However, we can get an idea of what is going on by slightly deforming the equation, for example,

In[·]:= `ContourPlot[u^2 - 4*v^2 - v^3 +`
`256/27 - 0.2 == 0, {u, -20,`
`20}, {v, -20, 20}, ColorFunction`
`-> (Red & )]`

Out[·]=

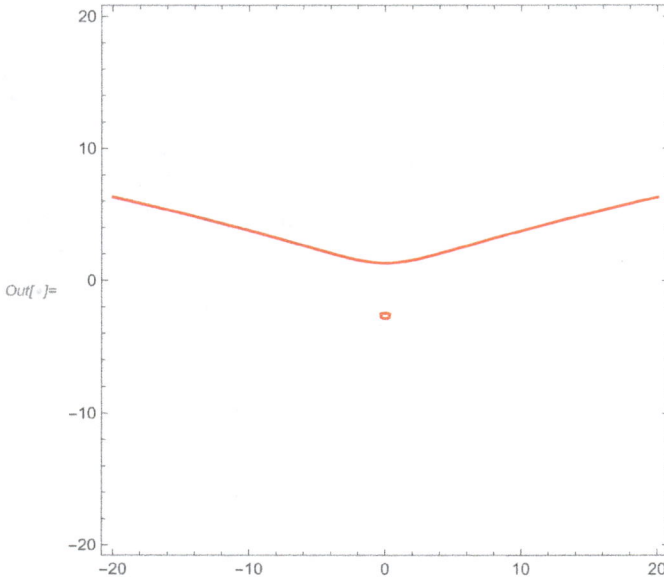

Figure 4.140

The real curve looks like this. The isolated point is the singular point. (Such a curve cannot be parametrized by an interval.) Another way to understand this phenomenon of an isolated point for this example is by using Plot3D:

In[·]:= `Plot3D[{u^2 - 4*v^2 - v^3 + 256/27, 0},`
`{u, -4, 4}, {v, -4, 4}]`

Out[·]=

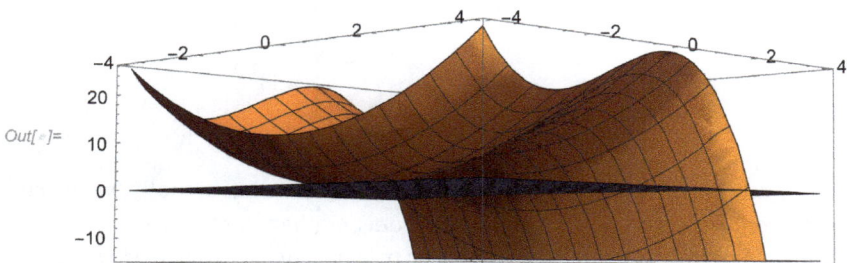

Figure 4.141

Various kinds of singularities exist also on surfaces. We will now define the tangent cone at a singular point. Let $p \in M$ be a point of an arbitrary subset M of $\mathbb{R}^n$. The tangent cone to M at p is defined as the set T_pM consisting of vectors $v \in \mathbb{R}^n$ such that for a sequence of points $x_n \in M \setminus \{p\}$ convergent to p we have $(x_n - p)/\|x_n - p\| \to v/\|v\|$. We also add the zero vector to T_pM. The tangent cone describes infinitesimal directions from the point p inside the set M.

Let us consider a surface given parametrically by

In[·]:= `fpar[u_, v_] := {v (v - 1) (v + 1),`
`        v (v + 1), u v}`

In[·]:= `ParametricPlot3D[fpar[u, v], {u,`
`        -3, 3}, {v, -3, 3}]`

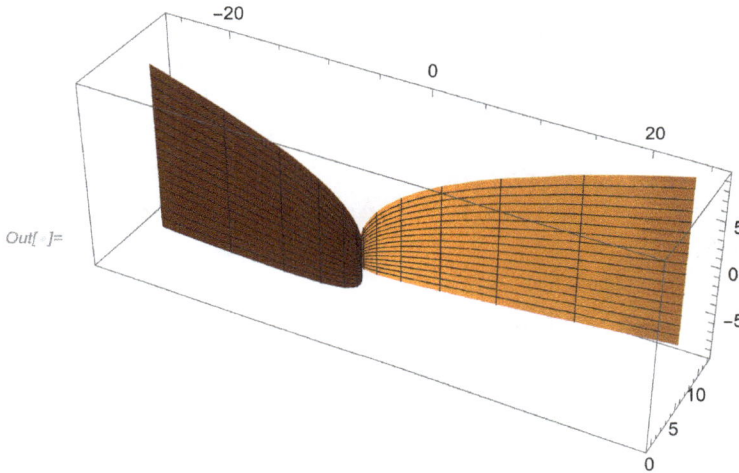

Figure 4.142

We see that the origin is a singular point on this surface. Let us find three linearly independent vectors which belong to the tangent cone at the origin. The first sequence of points on the surface convergent to the origin is

In[·]:= `x1[n_] := fpar[0, 1/n]`

In[·]:= `Limit[x1[n], n -> Infinity]`
Out[·]:= `{0, 0, 0}`

The second sequence of points on the surface convergent to the origin is

In[·]:= `x2[n_] := fpar[1/n, -1 + 1/n]`

In[·]:= `Limit[x2[n], n -> Infinity]`
Out[·]:= `{0, 0, 0}`

The third sequence of points on the surface convergent to the origin is

In[·]:= x3[n_] := fpar[0, -1 + 1/n]

In[·]:= Limit[x3[n], n -> Infinity]
Out[·]:= {0, 0, 0}

Now let us use the definition for the vectors in the tangent cone.

In[·]:= v1 = Limit[FullSimplify[(x1[n] -
        {0, 0, 0})/Norm[x1[n] - {0, 0, 0}]
        /. Abs[a_] -> a], n -> Infinity]
Out[·]:= {-(1/Sqrt[2]), 1/Sqrt[2], 0}

Similarly omitting the zeros in the numerator and the denominator

In[·]:= v2 = Limit[FullSimplify[x2[n]/
        Norm[x2[n]] /. Abs[a_] -> a],
        n -> Infinity]
Out[·]:= {Sqrt[2/3], -(1/Sqrt[6]), -(1/Sqrt[6])}

In[·]:= v3 = Limit[FullSimplify[x3[n]/
        Norm[x3[n]] /. Abs[a_] -> a],
        n -> Infinity]
Out[·]:= {2/Sqrt[5], -(1/Sqrt[5]), 0}

we see that the vectors are linearly independent:

In[·]:= Equal[Det[{v1, v2, v3}], 0]
Out[·]:= False

5 Elements of the theory of special functions

We have already encountered various special functions in Volumes 1, 2 [7, 8] and in this volume in the outputs of some expressions (e. g., result of integration). In this chapter we would like to touch briefly on the vast theory of special functions. This is not by all means a complete account of the theory since the "zoo" of special functions is very big and all functions have remarkable and intriguing properties. They also have many connections to other areas of mathematics, mathematical physics, number theory, and so on. We have chosen a couple of illustrative examples and refer the interested reader to more specialized books and the Documentation Center.

Classical special functions (for instance, Airy, Bessel, hypergeometric functions, and their confluent cases when singularities merge) satisfy linear ordinary differential equations (ODEs) and/or linear difference equations. For special values of the parameters, they may be expressed in terms of rational or other elementary functions. There is a vast body of literature on classical special functions. Many of them are implemented in Mathematica® as well (see guide/SpecialFunctions). Nonlinear special functions (for instance, the Painlevé transcendents which are solutions of certain nonlinear second-order differential equations known as the Painlevé equations) also have many remarkable properties. However, at present, they are not yet implemented in Mathematica®. In this chapter we shall discuss some aspects of the chosen special functions. Throughout this chapter we assume that both dependent and independent variables are complex unless explicitly stated otherwise. We consider a few topics from the theory of linear and nonlinear special functions and show how computations involving special functions can be done using Mathematica®. We shall use different reference books and various internet resources (like the Mathematical Functions Site [24], WolframMathWorld™ [25], NIST Digital Library of Mathematical Functions DLMF [18, 29], and others). We would also like to remark that there exist entities `MathematicalFuntion` and `MathematicalFunctionData` which return extensive information about built-in (classical and special) functions. For the section on linear special functions below, we shall use [12, 15] as our reference books.

5.1 Linear special functions

In many cases Mathematica® can compute singularities of elementary functions over the real or complex numbers:

```
In[·]:= FunctionSingularities[Log[x], x]
Out[·]:= x <= 0
```

```
In[·]:= FunctionSingularities[Log[x], x, Complexes]
Out[·]:= x == 0 || (Im[x] == 0 && Re[x] <= 0)
```

https://doi.org/10.1515/9783110774641-005

In[·]:= FunctionSingularities[Tan[I x], x, Complexes]
Out[·]:= Cosh[x] == 0

A generalization of the Taylor series for complex functions is the Laurent series. Mathematica® can compute Taylor and Laurent expansions with a finite principal part but when the function has essential singularities Mathematica®'s function Series returns the function itself:

In[·]:= Normal[Series[E^z, {z, 0, 2}]]
Out[·]:= 1 + z + z^2/2

In[·]:= Series[E^z/z, {z, 0, 2}] // Normal
Out[·]:= 1 + 1/z + z/2 + z^2/6

In[·]:= Normal[Series[E^(1/z), {z, 0, 3}]]
Out[·]:= E^(1/z)

The same is true for the function Residue:

In[·]:= Residue[1/z, {z, 0}]
Out[·]:= 1

In[·]:= Residue[E^(1/z), {z, Infinity}]
Out[·]:= -1

In[·]:= Residue[E^(1/z), {z, 0}]
Out[·]:= Residue[E^(1/z), {z, 0}]

Differential equations are usually very difficult to solve explicitly. However, the singularities of a solution to a linear ordinary differential equation can be read off from the coefficients of the equation, in contrast to solutions of nonlinear ODEs. It is sufficient to divide the equation by its leading coefficient (appearing in front of the highest derivative) and study the singularities of other coefficients, including those at infinity (which is usually done by using a change of variables).

An important technique in dealing with differential equations is change of variables. Let us make the substitution $z \to 1/x$ in the following linear ODE:

In[·]:= Simplify[z^2*D[y[z], {z, 2}] -
        (a + b - 1)*z*Derivative[1][y][z]
        + a*b*y[z] /. {y[z] -> y1[1/z],
        D[y[z] -> y1[1/z], z], D[y[z] ->
        y1[1/z], {z, 2}]} /. z -> 1/x]
Out[·]:= a*b*y1[x] + x*((1 + a + b)*
        Derivative[1][y1][x] +
        x*Derivative[2][y1][x])

From above we can easily see that $z = 0$ and $z = \infty$ are the only singularities of the solutions of this equation. Such singularities are said to be fixed since they depend on the equation and not on the particular solution. There is a further classification of singularities into regular and irregular ones [15].

For solutions of linear differential equations, the local behavior near regular singularities can be determined from the so-called indicial equation $r(r-1) + p_0 r + q_0 = 0$, where p_0 and q_0 are residues of certain expressions involving coefficients of the equation at singular points [15]. We have already seen that the following equation (and hence its solutions) has singularities at $z = 0$ and $z = \infty$.

```
In[·]:= EulerEq1 = z^2*D[y[z], {z, 2}] -
        (a + b - 1)*z*Derivative[1][y][z]
        + a*b*y[z];
```

```
In[·]:= P0coeff = Coefficient[EulerEq1,
          Derivative[2][y][z]]
Out[·]:= z^2
```

```
In[·]:= P1coeff = Coefficient[EulerEq1,
          Derivative[1][y][z]]
Out[·]:= -((-1 + a + b)*z)
```

```
In[·]:= P2coeff = Coefficient[EulerEq1,
          y[z]]
Out[·]:= a*b
```

```
In[·]:= p0 = Residue[P1coeff/P0coeff, {z, 0}]
Out[·]:= 1 - a - b
```

```
In[·]:= q0 = Residue[(z)*(P2coeff/
          P0coeff), {z, 0}]
Out[·]:= a*b
```

```
In[·]:= Solve[r (r - 1) + p0 r + q0 == 0, r]
Out[·]:= {{r -> a}, {r -> b}}
```

These are characteristic exponents at $z = 0$ and they mean that we can find local expansions of solutions near the singularity in the form of the Frobenius series $y(z) = z^a \sum_{k=0}^{\infty} c_{k,1} z^k$ and $y(z) = z^b \sum_{k=0}^{\infty} c_{k,2} z^k$. We additionally assume here that the difference of the local exponents is not an integer, otherwise some modifications of the series are needed. To find the coefficients $c_{k,1}$ and $c_{k,2}$ explicitly, we should use a finite number of terms of the formal expansion with a small k:

```
In[·]:= Collect[EulerEq1 /. {y[z] -> z^a*Sum[c[k, 1]
        *(z + 0)^k, {k, 0, 2}], D[y[z] ->
        z^a*Sum[c[k, 1]*(z + 0)^k, {k,
        0, 2}], z], D[y[z] -> z^a*Sum[c[k, 1]
        *(z + 0)^k, {k, 0, 2}], {z, 2}]},
        z^(_.)]
Out[·]:= z^(1 + a)*(c[1, 1] + a*c[1, 1] -
        b*c[1, 1]) + z^(2 + a)*(4*c[2, 1]
        + 2*a*c[2, 1] - 2*b*c[2, 1])
```

At this step one can either write a function which calculates coefficients (using, for instance, CoefficientList, Take, Solve, etc.), but usually it is more convenient to do this manually depending on the problem at hand since some special cases may be lost. Moreover, one needs to take a sufficient number of terms to make sure that the calculations are valid.

It is not difficult to see from the expansion above that the first few coefficients $c_{1,1}$, $c_{2,1}$ are zero. If we take more terms in the expansions, all subsequent coefficients are zero as well. To show this rigorously, one needs to derive a recurrence relation by hand. So we are left only with an expansion $y(z) = c_{0,1}z^a$, where $c_{0,1}$ is arbitrary. Similar computations can be done for the second local exponent $r = b$, and one also obtains a local expansion $y(z) = c_{0,2}z^b$, where $c_{0,2}$ is arbitrary. In our case we are lucky since the Euler equation can indeed be solved explicitly as follows:

```
In[·]:= DSolve[EulerEq1 == 0, y[z], z]
Out[·]:= {{y[z] -> z^a*C[1] + z^b*C[2]}}
```

We can also try to use the function AsymptoticDSolveValue, which should work in the case of both regular and irregular singularities, for instance,

```
In[·]:= AsymptoticDSolveValue[EulerEq1
        == 0, y[z], {z, 0, 3}]
Out[·]:= z^b*C[1] + z^a*C[2]
```

```
In[·]:= AsymptoticDSolveValue[{Derivative
        [2][y][x] + 2*y[x] == 0, y[0] ==
        1, Derivative[1][y][0] == 0},
        y[x], {x, 0, 3}]
Out[·]:= 1 - x^2
```

```
In[·]:= AsymptoticDSolveValue[Sin[x]*
        Derivative[1][y][x] + y[x] ==
        0, y[x], {x, 0, 2}]
Out[·]:= ((1 - x^2/12)*C[1])/x
```

In[·]:= AsymptoticDSolveValue[x^3*
　　　 Derivative[1][y][x] + y[x] == 0,
　　　 y[x], {x, 0, 2}]
Out[·]:= E^(1/(2*x^2))*C[1]

Mathematica® knows many expansions of special functions. For instance,

In[·]:= bessel = z^2*Derivative[2][y][z] +
　　　 z*Derivative[1][y][z] + (z^2 -
　　　 v^2)*y[z];

In[·]:= DSolve[bessel == 0, y[z], z]
Out[·]:= {{y[z] -> BesselJ[v, z]*C[1]
　　　 + BesselY[v, z]*C[2]}}

In[·]:= Normal[Series[BesselJ[v, z],
　　　 {z, 0, 2}]]
Out[·]:= z^v*(1/(2^v*Gamma[1 + v]) -
　　　 (2^(-2 - v)*z^2)/((1 + v)*
　　　 Gamma[1 + v]))

In[·]:= Simplify[Normal[Series[BesselY
　　　 [v, z], {z, 0, 1}]]]
Out[·]:= -(((z^v*Cos[Pi*v]*Gamma[-v])/2^v
　　　 + (2^v*Gamma[v])/z^v)/Pi)

From above we easily see that the solutions of the indicial equation (the characteristic exponents) at $z = 0$ are $r = \pm v$.

Another example is the Airy equation:

In[·]:= DSolve[Derivative[2][y][z] - z*y[z]
　　　 == 0, y[z], z]
Out[·]:= {{y[z] -> AiryAi[z]*C[1] +
　　　 AiryBi[z]*C[2]}}

In[·]:= Normal[Series[AiryAi[z], {z,
　　　 Infinity, 2}]]
Out[·]:= (-(5/(96*Sqrt[Pi]*z^(7/4))) +
　　　 1/(2*Sqrt[Pi]*z^(1/4)))/
　　　 E^((2*z^(3/2))/3)

We see that infinity is an irregular singularity for the Airy equation.

5.2 The Gauss hypergeometric function

The Gauss hypergeometric function appears when one integrates the simple expression below:

```
In[·]:= Integrate[((1 - x)^(c - b - 1)*
        x^(b - 1))/(1 - z*x)^a, {x, 0, 1}]
```

Out[·]= Gamma[b] Gamma[-b + c] Hypergeometric2F1Regularized[a, b, c, z] if Re[b] < Re[c] && Re[b] > 0 && (Re[z] ≤ 1 || z ∉ ℝ)

Figure 5.1

In particular cases this function reduces to elementary functions:

```
In[·]:= Hypergeometric2F1[1, b, b, z]
Out[·]:= 1/(1 - z)
```

```
In[·]:= z Hypergeometric2F1[1, 1, 2, -z]
Out[·]:= Log[1 + z]
```

```
In[·]:= Hypergeometric2F1[x, -x, 1/2, 1]
Out[·]:= Cos[Pi*x]
```

Moreover, $_2F_1(a, b, c, z)$ also solves the following linear second-order ODE:

$$z(z - 1)y''(z) + ((a + b + 1)z - c)y'(z) + aby(z) = 0.$$

One more way to define the hypergeometric function is by using the following series:

```
In[·]:= Normal[Series[Hypergeometric2F1
        [a, b, c, z], {z, 0, 2}]]
Out[·]:= 1 + (a*b*z)/c + (a*(1 + a)*b*
        (1 + b)*z^2)/(2*c*(1 + c))
```

Let us prove the so-called differentiation formula in Mathematica®:

```
In[·]:= GHE[a_, b_, c_, y_, z_] := a*b*y +
        (z*(1 + a + b) - c)*D[y, z] +
        (-1 + z)*z*D[y, {z, 2}]
```

```
In[·]:= Solve[GHE[a, b, c, y[z], z] ==
        0, Derivative[2][y][z]];
```

```
In[·]:= Together[GHE[a + 1, b + 1, c + 1,
        (c/(a*b))*Derivative[1][y][z], z]
        //. {%[[1]][[1]], D[%[[1]][[1]], z]}]
Out[·]:= 0
```

The Gauss hypergeometric function satisfies many identities. Most of them are implemented in Mathematica®. For instance, the parameters a and b can be interchanged:

```
In[·]:= Hypergeometric2F1[b, a, c, z]
Out[·]:= Hypergeometric2F1[a, b, c, z]
```

In[·]:= Hypergeometric2F1[a, b, c, z] ==
    Hypergeometric2F1[b, a, c, z]
Out[·]:= True

It is useful to know how to prove some of these identities. Let us consider the example of a quadratic transformation for $_2F_1$:

$$_2F_1(a, b, (a + b + 1)/2, z) = {_2F_1}(a/2, b/2, (a + b + 1)/2, 1 - 4(z - 1/2)^2).$$

The first method to show that this formula is correct is by expanding both sides (or their difference) into a power series around the point $z = 0$. We can use only a finite number of terms:

In[·]:= Normal[Series[FullSimplify[
    Hypergeometric2F1[a, b, (a + b + 1)/2, z]
    - Hypergeometric2F1[a/2, b/2, (a + b + 1)/2,
    1 - 4*(z - 1/2)^2]], {z, 0, 10}]]
Out[·]:= 0

Another, and more rigorous, method is to use differential equations. We can make a change of variables in the equation and simplify.

In[·]:= Solve[1 - 4*(z - 1/2)^2 == t, z]
Out[·]:= {{z -> (1/2)*(1 - Sqrt[1 - t])},
    {z -> (1/2)*(1 + Sqrt[1 - t])}}

In[·]:= GHE[a/2, b/2, (a + b + 1)/2, y[t], t]
    /. {y[t] -> y1[(1/2)*(1 + Sqrt[1 - t])],
    D[y[t] -> y1[(1/2)*(1 + Sqrt[1 - t])], t],
    D[y[t] -> y1[(1/2)*(1 + Sqrt[1 - t])],
    {t, 2}]};

In[·]:= Simplify[PowerExpand[Together[% /.
    t -> 1 - 4*(z - 1/2)^2]]];

In[·]:= Simplify[% /. Solve[GHE[a, b,
    (a + b + 1)/2, y1[z], z] == 0,
    Derivative[2][y1][z]][[1]]]
Out[·]:= 0

5.3 Legendre polynomials

There are many well-known orthogonal polynomials. Some of them are implemented in Mathematica®. However, if one wants to study semiclassical orthogonal polynomials, one has to implement them oneself.

Orthogonal polynomials satisfy a three-term recurrence relation [4]. Classical orthogonal polynomials (e. g., Jacobi polynomials or, more generally, polynomials in the Askey–Wilson scheme) have many interesting properties. It is also remarkable that the recurrence coefficients of certain semiclassical orthogonal polynomials with specific weights turn up as classical solutions of nonlinear Painlevé equations discussed in one of the next subsections [6, 17].

Let us define Legendre polynomials by

In[·]:= p[n_][x_] := (1/(2^n*n!))*D[(x^2 - 1)^n, {x, n}]

This definition of polynomials coincides with the built-in function LegendreP:

In[·]:= Table[p[n][x] - LegendreP[n, x],
        {n, 0, 5}] // Simplify
Out[·]:= {0, 0, 0, 0, 0, 0}

For general n, Mathematica® returns the answer in terms of a DifferenceRoot:

In[·]:= p[n][x]

$$Out[\cdot]= \; 2^{-n} \; \boxed{\triangle \; y[2+n]} \; [n]$$

Figure 5.2

In InputForm we have

In[·]:= p[n][x]

Out[·]= DifferenceRoot[Function[{y, n}, {(n - 2*n)*y[n] - 2*(-1 - n + n)*x*y[1 + n] +
        (2 + n)*(-1 + x)*(1 + x)*y[2 + n] = 0, y[0] = (-1 + x^2)^n, y[1] = 2*n*x*(-1 + x^2)^(-1 + n)}]][
    n]/2^n

Figure 5.3

The same answer is returned if we use FunctionExpand:

In[·]:= FunctionExpand[p[n][x]];

or the same function with assumptions

In[·]:= FunctionExpand[p[n][x], Assumptions
        -> {Element[n, PositiveIntegers], x > 0}];

However, DiferenceRootReduce does not return this DifferenceRoot object (it should!):

In[·]:= DifferenceRootReduce[LegendreP[n, x]]
Out[·]:= LegendreP[n, x]

In[·]:= DifferenceRootReduce[LegendreP[5, x]]

Out[ ]= DifferenceRoot[Function[{y, n}, {(-8 - 120*n - 420*n^2 - 560*n^3 - 315*n^4 - 63*n^5)*y[n] +
    (15*n - 70*n^3 + 63*n^5)*y[1 + n] == 0, y[-1] == -1, y[1] == 1}]][x]

Figure 5.4

We can check that Legendre polynomials are orthogonal on the interval $x \in [-1, 1]$:

```
In[·]:= Together[Table[Integrate[p[k][x]*
           p[j][x], {x, -1, 1}], {j, 0, 3},
           {k, 0, 3}]]
Out[·]:= {{2, 0, 0, 0}, {0, 2/3, 0, 0},
          {0, 0, 2/5, 0}, {0, 0, 0, 2/7}}
```

We can see that the hypergeometric function for special values of parameters coincide with them:

```
In[·]:= FullSimplify[Hypergeometric2F1[n + 1,
           -n, 1, (1 - x)/2] - LegendreP[n, x]]
Out[·]:= 0
```

Legendre polynomials are solutions of the following linear differential equation:

```
In[·]:= LEq[n_, y_, x_] := (1 - x^2)*
           D[y, {x, 2}] - 2*x*D[y, x] +
           n*(n + 1)*y

In[·]:= Table[LEq[n, p[n][x], x], {n, 0, 5}]
           // Together
Out[·]:= {0, 0, 0, 0, 0, 0}
```

We can next show that the Legendre differential equation becomes the hypergeometric equation after changing variables:

```
In[·]:= Solve[z == (1 - x)/2, x]
Out[·]:= {{x -> 1 - 2 z}}

In[·]:= Simplify[LEq[n, y[x], x] /. {y[x] ->
           y1[(1 - x)/2], D[y[x] -> y1[(1 - x)/2],
           x], D[y[x] -> y1[(1 - x)/2],
           {x, 2}]} /. x -> 1 - 2*z];

In[·]:= Together[% //. Solve[GHE[a, b,
           c, y1[z], z] == 0, Derivative[2]
           [y1][z]][[1]][[1]]];

In[·]:= Flatten[Simplify[CoefficientList[
           %, {Derivative[1][y1][z], y1[z], z}]]];
```

In[·]:= Solve[% == 0, {a, b, c}]
Out[·]:= {{a -> -n, b -> 1 + n, c-> 1},
        {a -> 1 + n, b -> -n, c -> 1}}

This gives us two sets of solutions for unknown parameters of the hypergeometric equation. Indeed, we also have

In[·]:= FullSimplify[Hypergeometric2F1[
        n + 1, -n, 1, (1 - x)/2] -
        LegendreP[n, x]]
Out[·]:= 0

This is no surprise since in the hypergeometric equation parameters a and b can be interchanged, as mentioned in the previous subsection.

A linear second-order differential equation with four regular singularities is the so-called Heun's equation. The solution of this equation, the Heun function, is implemented in Mathematica® as HeunG (ref/HeunG). It is in a sense a generalization of the Gauss hypergeometric function, but many properties are different. Both hypergeometric and Heun functions (and their confluences) are used widely in mathematics and mathematical physics.

5.4 Example: Discrete orthogonal polynomials

Let us find polynomials $\pi_k(x)$ of degree k ($k = 0, \ldots, n$), orthogonal with respect to the discrete m-point measure with the Dirac delta function $d\lambda(x) = \sum_{i=1}^{m} \delta(x - x_i)$. We assume that $(p(x), q(x))_{d\lambda(x)} = \int_{\mathbb{R}} p(x)q(x)d\lambda(x)$ is the inner product on the real line and $\|p(x)\|$ stands for the norm $\|p(x)\| = \sqrt{(p(x), p(x))_{d\lambda(x)}}$. The (monic) polynomials $\pi_k(x)$ orthogonal with respect to the measure can be defined via the moments $m_k = \int_{\mathbb{R}} x^k d\lambda(x)$ by the following formula (see, for instance, [9]):

$$
\pi_k(x) = \text{Det}\left[\begin{pmatrix} m_0 & m_1 & \cdots & m_k \\ m_1 & m_2 & \cdots & m_{k+1} \\ \vdots & \vdots & \ddots & \vdots \\ 1 & x & \cdots & x^k \end{pmatrix}\right] / \text{Det}[(m_{i+j})_{i,j=0}^n].
$$

Let us compute the first few moments:

In[·]:= m = 4;

We choose the points x_i as follows:

In[·]:= xi = {0, 1, 3, 5};

In[·]:= m0 = m;

```
In[·]:= m1 = Sum[xi[[i]], {i, 1, m}]
Out[·]:= 9
```

```
In[·]:= m2 = Sum[xi[[i]]^2, {i, 1, m}]
Out[·]:= 35
```

```
In[·]:= m3 = Sum[xi[[i]]^3, {i, 1, m}]
Out[·]:= 153
```

and then we can define the first few polynomials:

```
In[·]:= pol0[x_] := 1
```

```
In[·]:= pol1[x_] := Det[{{m0, m1}, {1, x}}]/m0
```

```
In[·]:= pol1[x]
Out[·]:= (1/4)*(-9 + 4*x)
```

```
In[·]:= pol2[x_] := Det[{{m0, m1, m2},
        {m1, m2, m3}, {1, x, x^2}}]/
        Det[{{m0, m1}, {m1, m2}}]
```

```
In[·]:= pol2[x]
Out[·]:= (1/59)*(152 - 297*x + 59*x^2)
```

One can easily write a short function to generate orthogonal polynomials (here we use new notation μ_k for moments):

```
In[·]:= polnew[x_, k_ /; k >= 2] :=
        Det[Append[Drop[Table[mu[i + j],
        {i, 0, k}, {j, 0, k}], -1],
        Table[x^i, {i, 0, k}]]]/
        Det[Table[mu[i + j], {i, 0,
        k - 1}, {j, 0, k - 1}]]
```

It can be easily verified that our definition is correct, for instance, for $k = 2$:

```
In[·]:= polnew[x, 2] /. {mu[0] -> m0, mu[1]
        -> m1, mu[2] -> m2, mu[3] -> m3}
Out[·]:= (1/59)*(152 - 297*x + 59*x^2)
```

We can also check orthogonality of polynomials, i. e., $\sum_{i=1}^{m} \pi_k(x_i)\pi_j(x_i) = 0$ for $k \neq j$:

```
In[·]:= {Sum[pol0[xi[[i]]]*pol1[xi[[i]]],
        {i, 1, m}], Sum[pol0[xi[[i]]]*pol2
        [xi[[i]]], {i, 1, m}], Sum[pol1
        [xi[[i]]]*pol2[xi[[i]]], {i, 1, m}]}
Out[·]:= {0, 0, 0}
```

5.5 Nonlinear special functions and Painlevé transcendents

Solutions of nonlinear differential equations may have singularities which depend on initial conditions. For instance, we see that $z = -$const is a movable pole for the solution of a simple nonlinear first-order differential equation:

```
In[·]:= DSolve[Derivative[1][y][z] ==
          y[z]^2, y[z], z]
Out[·]:= {{y[z] -> 1/(-z - C[1])}}
```

If we impose initial conditions, then either $z = 1$ or $z = 1/2$ below is a pole:

```
In[·]:= DSolve[{Derivative[1][y][z] ==
          y[z]^2, y[0] == 1}, y[z], z]
Out[·]:= {{y[z] -> 1/(1 - z)}}
```

```
In[·]:= DSolve[{Derivative[1][y][z] ==
          y[z]^2, y[0] == 2}, y[z], z]
Out[·]:= {{y[z] -> -(2/(-1 + 2*z))}}
```

If we increase the degree of y in the right-hand side of the equation above then solutions have the so-called movable algebraic singularities:

```
In[·]:= DSolve[Derivative[1][y][z] ==
          y[z]^3, y[z], z]
Out[·]:= {{y[z] -> -(1/(Sqrt[2]*Sqrt[-z - C[1]]))},
          {y[z] -> 1/(Sqrt[2]*Sqrt[-z - C[1]])}}
```

Solutions are, therefore, not meromorphic, they are multivalued in the complex plane. We do not capture such singularities by using `AsymptoticDSolveValue` since **Mathematica**® returns only holomorphic expansions:

```
In[·]:= AsymptoticDSolveValue[Derivative[1][y][z]
          == y[z]^2, y[z], {z, z0, 1}]
Out[·]:= C[1] + (z - z0)*C[1]^2
```

Moreover, for some equations **Mathematica**® does not return any expansion at infinity:

```
In[·]:= AsymptoticDSolveValue[Derivative[1][y][z]
          == y[z]^2, y[z], {z, Infinity, 3}]
Out[·]:= -(1/z) + C[1]/z^2 - C[1]^2/z^3
```

```
In[·]:= AsymptoticDSolveValue[Derivative[2][y][z]
          == 2*y[z]^3 + z*y[z] + a, y[z],
          {z, Infinity, 2}]
Out[·]:= AsymptoticDSolveValue[Derivative[2][y][z]
          == a + z*y[z] + 2*y[z]^3, y[z],
          {z, Infinity, 2}]
```

In some cases Mathematica® can integrate nonlinear differential equations in terms of elementary or special functions, for instance, in terms of elliptic functions as in the example below, but in general it cannot solve an arbitrary nonlinear equation explicitly.

```
In[·]:= DSolve[Derivative[2][y][z] ==
        6*y[z]^2 + 1, y[z], z]
Out[·]:= {{y[z] -> WeierstrassP[z + C[1],
        {-2, C[2]}]}}
```

The six Painlevé equations are special nonlinear second-order differential equations whose solutions have no movable algebraic singularities [18]. Their numerical study is very complicated due to the presence of movable poles. They have a lot of applications in modern mathematics and mathematical physics. For instance, they arise as similarity reductions of certain partial differential equations (the modified Korteweg–de Vries equation in the following example):

```
In[·]:= Factor[D[v[t, x], t] - 6*v[t, x]^2*
        D[v[t, x], x] + D[v[t, x], {x, 3}] //.
        {v[t, x] -> w[x/(3*t)^3^(-1)]/
        (3*t)^3^(-1), D[v[t, x] -> w[x/(3*t)^
        3^(-1)]/(3*t)^3^(-1), t], D[v[t, x] ->
        w[x/(3*t)^3^(-1)]/(3*t)^3^(-1), x],
        D[v[t, x] -> w[x/(3*t)^3^(-1)]/(3*t)
        ^3^(-1), {x, 3}]} /. x ->
        z*3^(1/3)*t^(1/3)]
Out[·]:= -((w[z] + z*Derivative[1][w][z] +
        6*w[z]^2*Derivative[1][w][z] -
        Derivative[3][w][z])/(3*3^(1/3)*t^(4/3)))
```

One can easily show that the second Painlevé equation

$$w''(z) = 2w(z)^3 + zw(z) + \alpha,$$

where α is a complex parameter, solves this equation:

```
In[·]:= Together[% /. {Derivative[2][w][z] ->
        2*w[z]^3 + z*w[z] + a, D[Derivative[2]
        [w][z] -> 2*w[z]^3 + z*w[z] + a, z]}]
Out[·]:= 0
```

The second Painlevé equation can be written in a Hamiltonian form with a polynomial Hamiltonian:

```
In[·]:= H2 = (1/2)*p[z]^2 - (q[z]^2 +
        (1/2)*z)*p[z] - (a + 1/2)*q[z];
```

```
In[·]:= {D[q[z], z] == D[H2, p[z]],
        D[p[z], z] == -D[H2, q[z]]};
```

```
In[·]:= Solve[%[[1]], p[z]];
```

```
In[·]:= Simplify[%% /. %[[1]] /.
        D[%[[1]], z]]
```

```
Out[·]:= {True, a + z*q[z] + 2*q[z]^3 ==
         Derivative[2][q][z]}
```

We see that for the function q we have the second Painlevé equation.

In order to find the so-called σ-form of the second Painlevé equation, one should eliminate variables p and q using the Hamiltonian system above. Let

```
In[·]:= s[z] == H2;
```

```
In[·]:= Simplify[D[%, z] //. {Derivative[1]
        [q][z] -> -(z/2) + p[z] - q[z]^2,
        Derivative[1][p][z] -> 1/2 + a
        + 2*p[z]*q[z]}]
```

```
Out[·]:= p[z] + 2*Derivative[1][s][z] == 0
```

```
In[·]:= Simplify[D[%, z] /. {Derivative[1]
        [q][z] -> -(z/2) + p[z] - q[z]^2,
        Derivative[1][p][z] -> 1/2 + a
        + 2*p[z]*q[z]}]
```

```
Out[·]:= 1 + 2*a + 4*p[z]*q[z] + 4*
         Derivative[2][s][z] == 0
```

```
In[·]:= Solve[p[z] + 2*Derivative[1][s][z]
        == 0, p[z]]
```

```
Out[·]:= {{p[z] -> -2*Derivative[1][s][z]}}
```

```
In[·]:= Simplify[{1 + 2*a + 4*p[z]*q[z] +
        4*Derivative[2][s][z], p[z] +
        2*Derivative[1][s][z], s[z] -
        (p[z]^2/2 - (1/2 + a)*q[z] -
        p[z]*(z/2 + q[z]^2))} //.
        p[z] -> -2*Derivative[1][s][z]];
```

```
In[·]:= Simplify[Resultant[%[[1]], %[[3]]],
        q[z]]]
```

```
Out[·]:= 2*Derivative[1][s][z]*(1 + 4*a +
         4*a^2 + 32*s[z]*Derivative[1][s][z] -
         32*z*Derivative[1][s][z]^2 -
         64*Derivative[1][s][z]^3 -
         16*Derivative[2][s][z]^2)
```

Lower-order equations may be solutions of higher-order equations, for instance,

```
In[·]:= Simplify[Derivative[2][y][z] ==
        Derivative[1][y][z] + 2*y[z]*
        Derivative[1][y][z] //.
        {Derivative[1][y][z] -> y[z] +
        y[z]^2, D[Derivative[1][y]
        [z] -> y[z] + y[z]^2, z]}]
Out[·]:= True
```

Let us show how to find Riccati solutions for special values of the parameter α of the second Painlevé equation:

```
In[·]:= P2[z_, w_, a_] := D[w, {z, 2}] -
        (2*w^3 + z*w + a)
```

```
In[·]:= CoefficientList[P2[z, w[z], a]
        //. {Derivative[1][w][z] ->
        a[z]*w[z]^2 + b[z]*w[z] + c[z],
        D[Derivative[1][w][z] ->
        a[z]*w[z]^2 + b[z]*w[z] +
        c[z], z]}, w[z]];
```

```
In[·]:= % //. {a[z] -> e, D[a[z] -> e, z],
        e^2 -> 1, b[z] -> 0, D[b[z] -> 0,
        z]}
Out[·]:= {-a + Derivative[1][c][z],
        -z + 2*e*c[z], 0, 0}
```

```
In[·]:= Together[% //. {c[z] -> z/(2*e),
        D[c[z] -> z/(2*e), z]}]
Out[·]:= {(1 - 2*a*e)/(2*e), 0, 0, 0}
```

The value of the parameter is

```
In[·]:= Solve[%[[1]] == 0, a]
Out[·]:= {{a -> 1/(2*e)}}
```

and the Riccati equation is

```
In[·]:= Derivative[1][w][z] - (a[z]*w[z]^2
        + b[z]*w[z] + c[z]) /. {c[z] ->
        z/(2*e), a[z] -> e, b[z] -> 0}
Out[·]:= -(z/(2*e)) - e*w[z]^2 +
        Derivative[1][w][z]
```

This equation (we have omitted equal to zero) can be linearized to give the Airy equation:

```
In[·]:= Numerator[Together[-(z/(2*e)) -
        e*w[z]^2 + Derivative[1][w][z] /.
        {w[z] -> (-e^(-1))*(Derivative[1]
        [y][z]/y[z]), D[w[z] -> (-e^(-1))
        *(Derivative[1][y][z]/y[z]), z]}]]
Out[·]:= (-z)*y[z] - 2*Derivative[2][y][z]
```

One of the most useful properties of the second Painlevé equation (and of all the other six Painlevé equations except for the first one) is the existence of the groups of Bäcklund transformations isomorphic to certain affine Weyl groups. Such transformations relate solutions with different values of parameters.

```
In[·]:= subst = Solve[-a - z*w[z] -
        2*w[z]^3 + Derivative[2][w][z]
        == 0, Derivative[2][w][z]][[1]];
```

```
In[·]:= Together[P2[z, -w[z] - (2*a + 1)/
        (2*w[z]^2 + 2*Derivative[1][w][z]
        + z), a + 1] /. D[subst, z]
        /. subst]
Out[·]:= 0
```

As mentioned above, generic solutions of the second Painlevé equation are meromorphic functions. The point $z = \infty$ is a fixed singularity. Generic solutions have an infinite number of simple poles accumulating at the fixed essential singularity $z = \infty$. For special values of parameters, the second Painlevé equation has rational solutions. Using the Bäcklund transformation above, we, for instance, can find how a given rational solution is transformed to another rational solution. Indeed, solution $w = -1/z$ with $\alpha = 1$ is transformed to another rational solution with $\alpha = 2$:

```
In[·]:= P2[z, -1/z, 1]
Out[·]:= 0
```

```
In[·]:= Together[-w[z] - (2*a + 1)/
        (2*w[z]^2 + 2*Derivative[1][w][z]
        + z) /. {w[z] -> -z^(-1),
        D[w[z] -> -z^(-1), z], a -> 1}]
Out[·]:= -((2*(-2 + z^3))/(z*(4 + z^3)))
```

```
In[·]:= Together[P2[z, %, 2]]
Out[·]:= 0
```

The Bäcklund transformation can be implemented also as follows:

```
In[·]:= Function[{w, a}, {Together[-w -
        (2*a + 1)/(2*w^2 + 2*D[w, z] +
        z)], a + 1}][-z^(-1), 1]
Out[·]:= {-((2*(-2 + z^3))/(z*(4 + z^3))), 2}
```

Let us apply the Bäcklund transformation several times and visualize the roots of the denominator of the obtained (complicated) rational function.

```
In[·]:= BTr[{w_, a_}] := Together[{-w -
        (2*a + 1)/(2*w^2 + 2*D[w, z] + z),
        a + 1}]
```

```
In[·]:= Denominator[Nest[BTr, {-z^(-1),
        1}, 20][[1]]];
```

```
In[·]:= p20 = ReIm[z /. NSolve[% == 0, z]];
```

```
In[·]:= Graphics[Point[p20]]
```

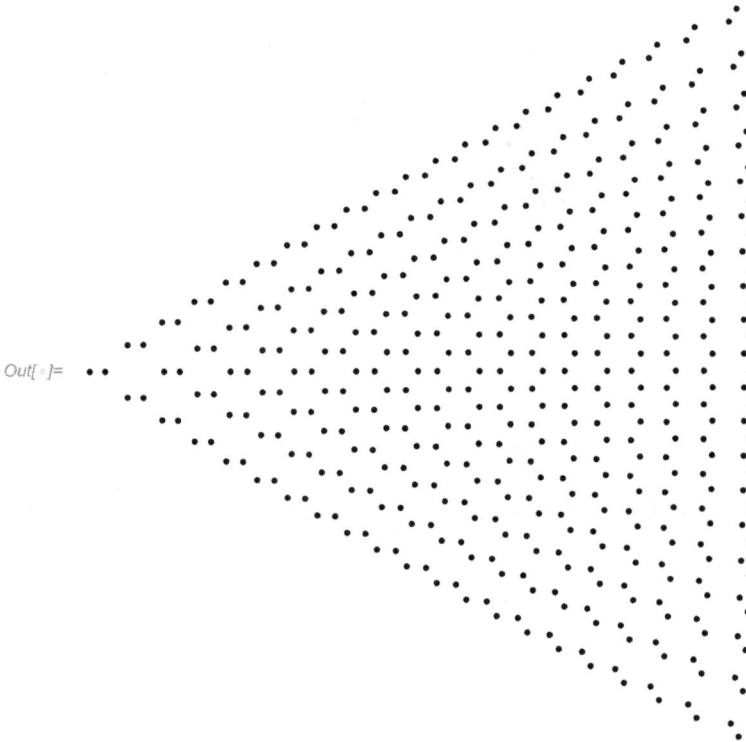

Out[·]=

Figure 5.5

We observe that the roots are located in some patterns in the complex plane. Moreover, there is a certain interlacing property if we compute the roots of the denominator of the rational function when one more Bäcklund transformation is applied:

```
In[·]:= Denominator[Nest[BTr, {-z^(-1), 1}, 21][[1]]];
```

```
In[·]:= p21 = ReIm[z /. NSolve[% == 0, z]];
```

In[·]:= Show[Graphics[{Red, Point[p20], Blue, Point[p21]}]]

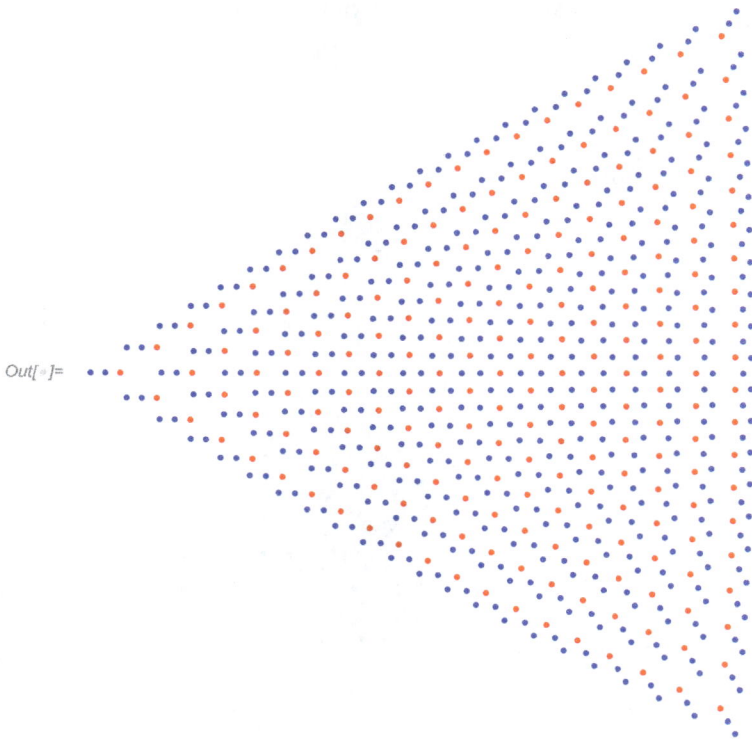

Out[·]=

Figure 5.6

More information about Painlevé transcendents and further references can be found in [18].

6 Elliptic functions

The development of the theory of elliptic functions goes back to the works of Gauss, Jacobi, Abel, Weierstrass, and many others [11]. Elliptic functions appear in different contexts. They are related to inversion of certain integrals, they can be defined as solutions of certain nonlinear differential equations, and they can be defined by certain series. They found applications in differential geometry of curves and surfaces, number theory, and many other fields of modern mathematics.

An elliptic function is a meromorphic doubly periodic function. It has only poles in the complex plane. A meromorphic function can be defined as a ratio of two entire functions. Elliptic functions have many remarkable properties, for instance, algebraic addition theorems. Addition theorems, in particular, can be used to extend the functions from real to complex variables. Also, the inverse statement due to Weierstrass holds: if a meromorphic function has an algebraic addition theorem and it is not rational with respect to t or $\exp(\alpha t)$, $\alpha \in \mathbb{C}$, then the function is elliptic. The theory of elliptic functions is also closely related to the theory of elliptic curves [5] and, more generally, of Riemann surfaces. One can find useful visualizations in [31]. Elliptic functions have many other remarkable properties and identities. We shall consider some of them in detail.

We remark that extensive information on built-in elliptic functions can be found in the Wolfram Documentation by typing guide/EllipticFunctions or tutorial/MathematicalFunctions #7229. In Mathematica® 12.3 elliptic functions and integrals were updated.

6.1 The generalized sine function

The usual trigonometric functions cos and sin can be defined as coordinates of a point on a unit circle, so the following identity holds: $\cos^2 t + \sin^2 t = 1$. The independent variable t can be taken as the length of a certain arc of the circle. It can also be taken as the double (signed) area of a certain curvelinear triangle [13]. In fact, if we define only one function, for instance, sin, then the other can be defined by $\cos t = \sin(\pi/2 - t)$. There are many trigonometric identities which are usually learnt at school. The most useful one is the addition formula, which allows one to express $\sin(a + b)$ or $\cos(a + b)$ in terms of the values of sin and cos at a and b. However, one can start with a different curve, a hyperbola or Bernoulli's lemniscate, and define functions with similar properties [13]. For instance, the hyperbolic functions are related to the hyperbola. Although the following two curves look very differently, the functions related to them, namely sine, hyperbolic sine (denoted by either sinh or sh), and also the lemniscate sine (usually denoted by sl and related to the lemniscate, see below), have many similar properties.

https://doi.org/10.1515/9783110774641-006

In[·]:= `ContourPlot[x^2 + y^2 == 1, {x, -1, 1},`
`{y, -1, 1}]`

Out[·]=

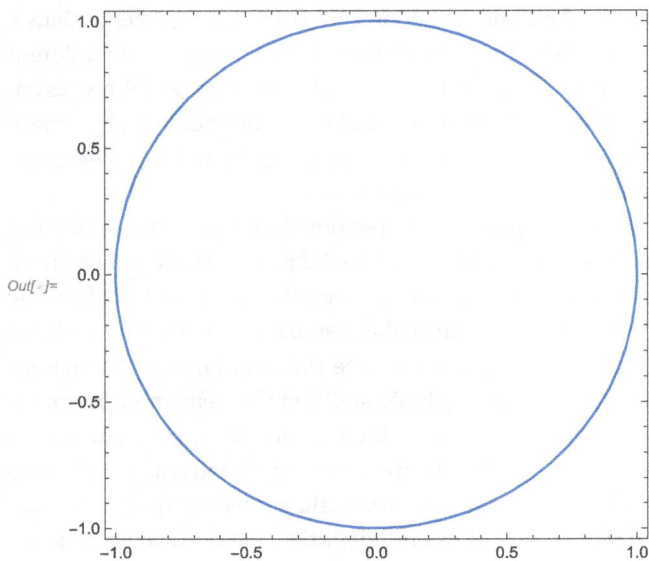

Figure 6.1

In[·]:= `ContourPlot[x^2 - y^2 == 1, {x, -10, 10},`
`{y, -10, 10}]`

Out[·]=

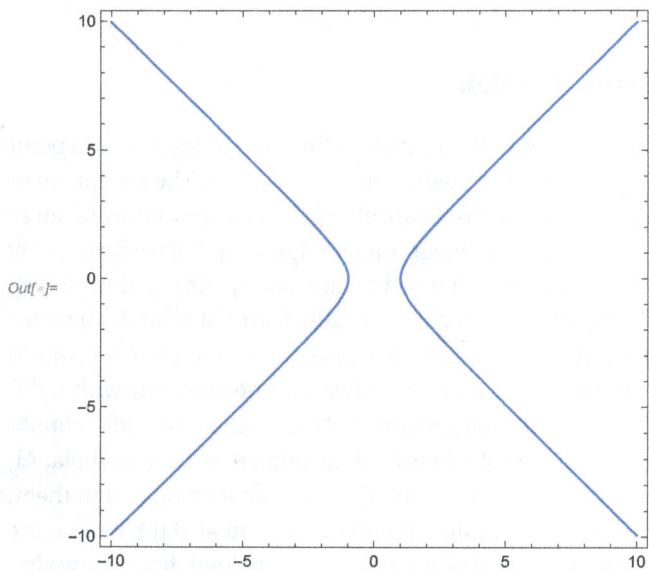

Figure 6.2

The lemniscate can be defined as a curve with the property that the product of distances to two fixed points is constant. We take $(1/2, 1/2)$ and $(-1/2, -1/2)$ as the fixed points and $1/2$ as the constant.

In[·]:= FullSimplify[Sqrt[(x - 1/2)^2 + (y - 1/2)^2]*
  Sqrt[(x + 1/2)^2 + (y + 1/2)^2] == 1/2];

Squaring both sides, we get

In[·]:= Simplify[%[[1]]^2 == %[[2]]^2]
Out[·]:= (x^2 + y^2)^2 == 2*x*y

This gives a fourth-degree algebraic curve.

In[·]:= ContourPlot[(x^2 + y^2)^2 == 2*x*y,
  {x, -1, 1}, {y, -1, 1}]

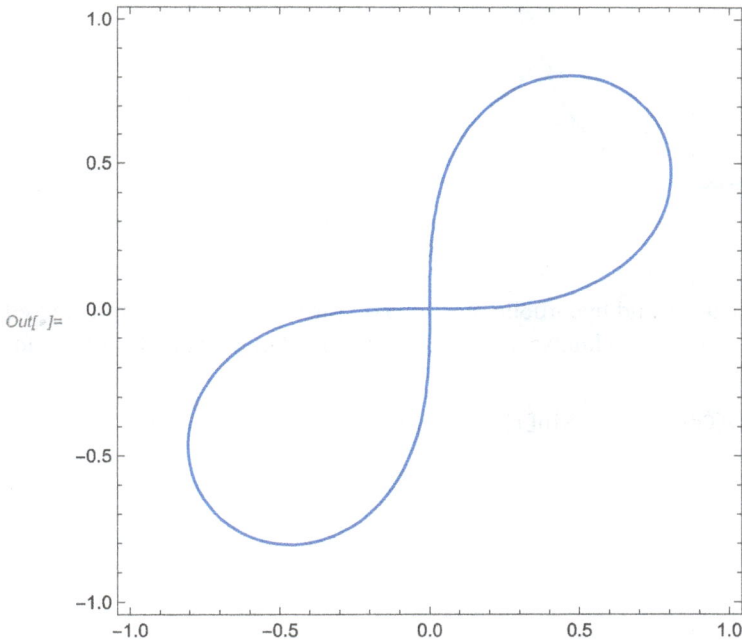

Figure 6.3

In polar coordinates we get

In[·]:= %% /. {x -> r Cos[t], y -> r Sin[t]} // Simplify
Out[·]:= r^3 == r*Sin[2*t]

so, by canceling r on both sides, we get an equation in polar coordinates $r^2 = \sin(2t)$. Mathematica® can draw curves given in polar coordinates:

In[·]:= PolarPlot[Sqrt[Sin[2*t]], {t, 0, 2*Pi}]

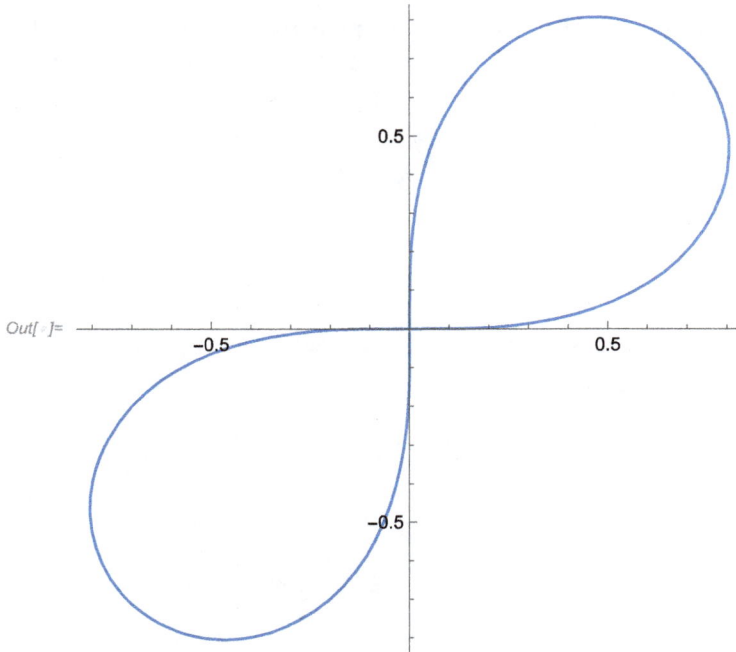

Figure 6.4

The usual trigonometric and hyperbolic functions on the unit circle are implemented in Mathematica®, which also knows many trigonometric identities and formulas, for instance,

In[·]:= Simplify[Cos[t]^2 + Sin[t]^2 == 1]
Out[·]:= True

In[·]:= Sin[Pi/2 - t]
Out[·]:= Cos[t]

In[·]:= Sin[-t]
Out[·]:= -Sin[t]

In[·]:= Sin[a + b] // TrigExpand
Out[·]:= Cos[b] Sin[a] + Cos[a] Sin[b]

In[·]:= Cos[a + b] // TrigExpand
Out[·]:= Cos[a] Cos[b] - Sin[a] Sin[b]

In[·]:= Sin[2 t] // TrigExpand
Out[·]:= 2 Cos[t] Sin[t]

In[·]:= Sin[Pi/2]
Out[·]:= 1

Mathematica® can solve simple equations:

In[·]:= Solve[Sin[t] == 1/2, t]

Out[·]= $\left\{\left\{t \to \boxed{\dfrac{\pi}{6} + 2\,\pi\,c_1 \text{ if } c_1 \in \mathbb{Z}}\right\}, \left\{t \to \boxed{\dfrac{5\,\pi}{6} + 2\,\pi\,c_1 \text{ if } c_1 \in \mathbb{Z}}\right\}\right\}$

Figure 6.5

The next equation has only nonreal solutions:

In[·]:= Solve[Sin[t] == 3, t]

Out[·]= $\left\{\left\{t \to \boxed{\pi - \text{ArcSin}[3] + 2\,\pi\,c_1 \text{ if } c_1 \in \mathbb{Z}}\right\}, \left\{t \to \boxed{\text{ArcSin}[3] + 2\,\pi\,c_1 \text{ if } c_1 \in \mathbb{Z}}\right\}\right\}$

Figure 6.6

If one wants only real solutions, one has to include the domain as the last argument.

In[·]:= Solve[Sin[t] == 3, t, Reals]
Out[·]:= {}

In[·]:= Solve[Sin[t] == 1, t, Reals]

Out[·]= $\left\{\left\{t \to \boxed{\dfrac{\pi}{2} + 2\,\pi\,c_1 \text{ if } c_1 \in \mathbb{Z}}\right\}\right\}$

Figure 6.7

Mathematica® can even solve certain trigonometric inequalities:

In[·]:= Reduce[Sin[2 t] > 0, t, Reals]
Out[·]:= Element[C[1], Integers] && Pi*C[1]
            < t < (1/2)*(Pi + 2*Pi*C[1])

For hyperbolic functions, we have

In[·]:= Simplify[Cosh[t]^2 - Sinh[t]^2 == 1]
Out[·]:= True

In[·]:= Sinh[-t]
Out[·]:= -Sinh[t]

In[·]:= Simplify[Cosh[t] - (E^t + E^(-t))/2]
Out[·]:= 0

We also have the following relations:

In[·]:= Sin[I*t]
Out[·]:= I*Sinh[t]

In[·]:= Sinh[I*t]
Out[·]:= I*Sin[t]

One can define the generalized sine function as the function inverse to the integral

In[·]:= Integrate[1/Sqrt[1 + m*z^2 + n*z^4],
      {z, 0, y}]

$$Out[\bullet]= -\left(\left(i \sqrt{2 + m y^2 - \sqrt{m^2 - 4n}\, y^2}\, \sqrt{2 + m y^2 + \sqrt{m^2 - 4n}\, y^2}\, \text{EllipticF}\left[i\, \text{ArcSinh}\left[\sqrt{2}\, \sqrt{\frac{n}{m + \sqrt{m^2 - 4n}}}\, y\right], -1 + \frac{m\left(m + \sqrt{m^2 - 4n}\right)}{2n}\right]\right)\right/$$

$$\left(2\sqrt{2}\, \sqrt{\frac{n}{m + \sqrt{m^2 - 4n}}}\, \sqrt{1 + m y^2 + n y^4}\right)\right)\quad \text{if}\quad \boxed{condition}\; —$$

Head: **And**
Byte count: **18016**
 Uniconize

Figure 6.8

In[·]:= f[m_, n_, y_] := Integrate[1/Sqrt[1 +
      m*z^2 + n*z^4], {z, 0, y}]

Here one assumes that $m^2 - 4n \neq 0$. As particular cases, we get the sine function

In[·]:= f[-1, 0, y]

Out[•]= **ArcSin[y]** if $-1 < \text{Re}[y] \leq 1\; ||\; y \notin \mathbb{R}$

Figure 6.9

In[·]:= InverseFunction[ArcSin][t]
Out[·]:= Sin[t]

the hyperbolic sine function

In[·]:= f[1, 0, y]

Out[•]= **ArcSinh[y]** if $-1 < \text{Im}[y] < 0\; ||\; 0 < \text{Im}[y] < 1\; ||\; \text{Re}[y] \neq 0$

Figure 6.10

In[·]:= InverseFunction[ArcSinh][t]
Out[·]:= Sinh[t]

Note that the following integral is divergent and Mathematica® does not evaluate it:

In[·]:= f[1, 0, Infinity];

The lemniscate sine is not implemented, but instead Mathematica® returns an expression in terms of the hypergeometric functions:

In[·]:= f[0, -1, y]

Out[●]= y Hypergeometric2F1$\left[\frac{1}{4}, \frac{1}{2}, \frac{5}{4}, y^4\right]$ if $0 \le \operatorname{Re}[y] < 1 \,\&\&\, \operatorname{Im}[y] == 0$

Figure 6.11

The plot of the inverse function resembles the plot of the sine function on a small interval:

In[·]:= Plot[{InverseFunction[#1*Hypergeometric2F1
        [1/4, 1/2, 5/4, #1^4] &][t], Sin[t]},
        {t, -1, 1}]

Out[ ]=

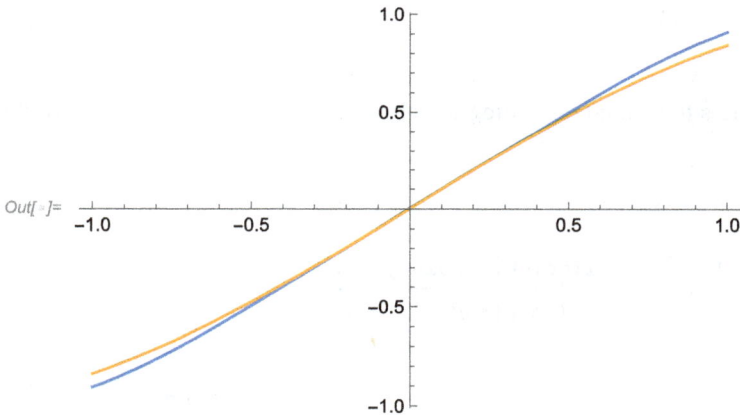

Figure 6.12

If one factorizes the expression under the square root in the integral above, one can get Jacobi's sn function as an inverse function.

In[·]:= Factor[Together[1 + m*z^2 + n*z^4 /.
        {m -> -1 - k^2, n -> k^2}]]
Out[·]:= (-1 + z)*(1 + z)*(-1 + k*z)*(1 + k*z)

It is implemented in Mathematica® as JacobiSN.

In[·]:= Plot[JacobiSN[t, 1/3], {t, -10, 10}]

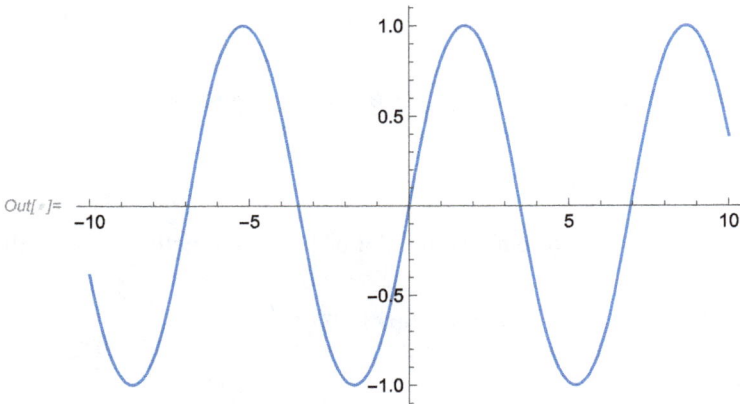

Figure 6.13

However, if one tries to evaluate the integral below, then **Mathematica**® does not do this immediately:

In[·]:= f[-1 - k^2, k^2, y]

Out[●]=
$$
\frac{\sqrt{1-y^2}\ \sqrt{1-k^2\,y^2}\ \text{EllipticF}\left[\text{ArcSin}[y],\,k^2\right]}{\sqrt{\left(-1+y^2\right)\times\left(-1+k^2\,y^2\right)}}
$$
if *condition* ▬

Head: **And**
Byte count: **23 088**

↗ Uniconize

Figure 6.14

When $y = 1$, we get a complete elliptic integral of the first kind $K(k^2)$, where k is the modulus. One needs to be a little careful with notation since in some books, for instance, in [13], the notation $K(k)$ is used:

In[·]:= f[-1 - k^2, k^2, 1]

Out[●]= $\text{EllipticK}\left[k^2\right]$ if $\text{Re}\left[k^2\right] \le 1\ ||\ k^2 \notin \mathbb{R}$

Figure 6.15

Elliptic K function also appears after integrating the following expression:

In[·]:= Integrate[1/Sqrt[1 - m*Sin[t]^2], {t,
        0, Pi/2}, Assumptions -> -1 < m < 1]
Out[·]:= EllipticK[m/(-1 + m)]/Sqrt[1 - m]

Mathematica® knows how to expand sn function into a power series with respect to the independent variable and with respect to the parameter:

In[·]:= Normal[Series[JacobiSN[z, k], {z, 0, 3}]]
Out[·]:= z + (1/6)*(-1 - k)*z^3

In[·]:= Normal[Series[JacobiSN[z, k], {k, 0, 1}]]
Out[·]:= Sin[z] + (1/8)*k*Cos[z]*(-2*z + Sin[2*z])

and how to integrate it

In[·]:= Integrate[JacobiSN[t, k], t]
Out[·]:= -(ArcTanh[Sqrt[k]*JacobiCD[t, k]]/Sqrt[k])

Also,

In[·]:= Limit[JacobiSN[z, k^2], k -> 0]
Out[·]:= Sin[z]

The inverse function is given by InverseJacobiSN. However, not all expressions are automatically evaluated in Mathematica® in comparison with sin and sinh above, for instance,

In[·]:= JacobiSN[I*t, 1/2]
Out[·]:= JacobiSN[I*t, 1/2]

In[·]:= JacobiSN[I, 1/2]
Out[·]:= JacobiSN[I, 1/2]

The following identity is known to be true but Mathematica® cannot verify it:

In[·]:= FullSimplify[JacobiSN[I*t, k^2] ==
        I*(JacobiSN[t, 1 - k^2]/Sqrt[1 -
        JacobiSN[t, 1 - k^2]^2])]

Let us introduce a function which represents the difference between the two sides:

In[·]:= func[t_, k_] := JacobiSN[I*t, k^2] -
        I*(JacobiSN[t, 1 - k^2]/Sqrt[1 -
        JacobiSN[t, 1 - k^2]^2])

We can easily check that the Taylor expansion at 0 with respect to the first variable is zero.

In[·]:= Normal[Series[func[t, 1/3], {t, 0, 10}]]
Out[·]:= 0

The same is true if we choose the Taylor expansion at other points. Mathematica® may give a warning that it is unable to decide whether some numeric quantities are equal to zero. A similar warning that the internal precision limit is reached appears if we try to substitute some (random) numerical values for *t* and *k* and use the function N with several digits precision. One needs to be also careful with branches of the square root.

Let us plot the real and imaginary parts.

In[·]:= Plot[ReIm[func[t, 1/2]], {t, 0, 1}]

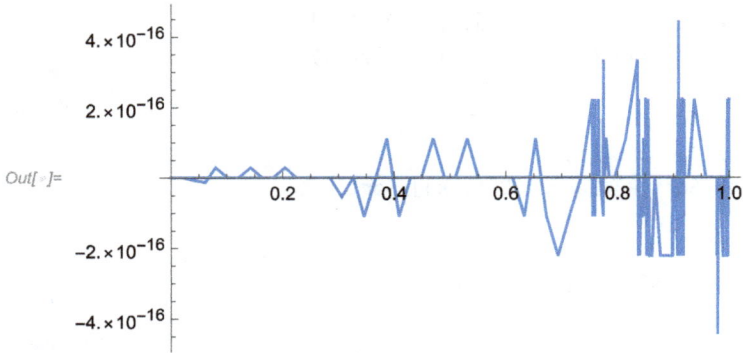

Figure 6.16

We see that the imaginary part is nonzero. This is also seen when we use the function ComplexPlot3D:

In[·]:= ComplexPlot3D[func[z, 1/2],
{z, -1 - I, 1 + I}]

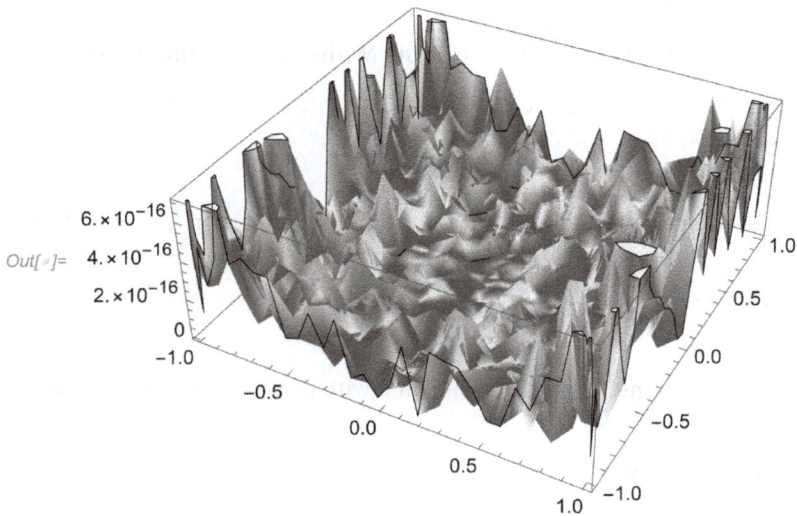

Figure 6.17

However, if we use Mathematica®'s ability to plot with arbitrary precision, by setting the option WorkingPrecision to some positive integer (even a small one), we see that both the real and imaginary parts come out as zero.

```
In[·]:= Plot[ReIm[func[t, 1/2]], {t, 0, 1},
         WorkingPrecision -> 5]
```

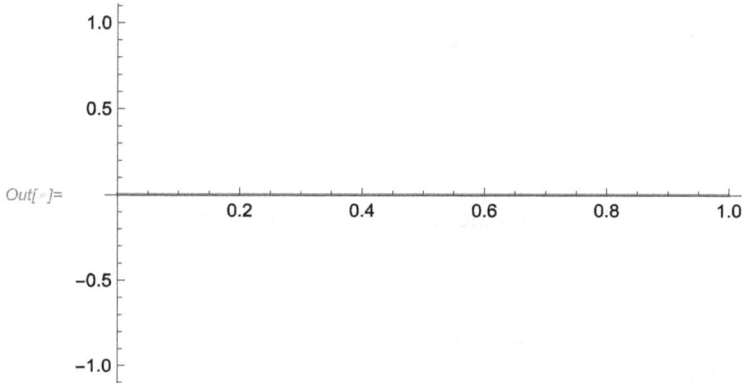

Figure 6.18

One can show that Jacobi's sn function has both zeros and poles (which are, in fact, periodic), for instance,

```
In[·]:= JacobiSN[2*EllipticK[k^2], k^2]
Out[·]:= 0
```

```
In[·]:= JacobiSN[2*I*EllipticK[1 - k^2], k^2]
Out[·]:= 0
```

```
In[·]:= JacobiSN[2*EllipticK[k^2] +
         3*I*EllipticK[1 - k^2], k^2]
Out[·]:= ComplexInfinity
```

However,

```
In[·]:= Solve[JacobiSN[t, k^2] == 0, t]
··· Solve: Inverse functions are being used by Solve, so some solutions may not
be found; use Reduce for complete solution information.
Out[·]:= {{t -> 0}}
```

```
In[·]:= Solve[JacobiSN[t, k^2] == 0, t,
         Method -> Reduce]
```

$Out[\bullet]= \left\{ \left\{ t \to 4\, c_1\, \text{EllipticK}\left[k^2\right] + 2\, i\, c_2\, \text{EllipticK}\left[1 - k^2\right] \text{ if } (c_1 \mid c_2) \in \mathbf{Z} \right\}, \right.$

$\left. \left\{ t \to 2\, \text{EllipticK}\left[k^2\right] + 4\, c_1\, \text{EllipticK}\left[k^2\right] + 2\, i\, c_2\, \text{EllipticK}\left[1 - k^2\right] \text{ if } (c_1 \mid c_2) \in \mathbf{Z} \right\} \right\}$

Figure 6.19

The function Reduce returns a similar answer:

$In[\cdot]:= \text{Reduce[JacobiSN[t, k\^2]} == 0, \text{t]}$

$Out[\bullet]= (c_1 \mid c_2) \in \mathbf{Z}\ \&\&$
$(t = 4\, c_1\, \text{EllipticK}\left[k^2\right] + 2\, i\, c_2\, \text{EllipticK}\left[1 - k^2\right] \mid\mid t = 2\, \text{EllipticK}\left[k^2\right] + 4\, c_1\, \text{EllipticK}\left[k^2\right] + 2\, i\, c_2\, \text{EllipticK}\left[1 - k^2\right])$

Figure 6.20

Let us denote the length of the lemniscate as 2ω. We can compute the value $\omega/2$ in Mathematica® by using the function ArcLength. Below we use the polar coordinates with $r = \sqrt{\sin(2t)}$ for t ranging from 0 to $\pi/4$:

```
In[·]:= Simplify[ArcLength[{Sqrt[Sin[2*t]]*
           Cos[t], Sqrt[Sin[2*t]]*Sin[t]},
           {t, 0, Pi/4}]]
Out[·]:= EllipticK[1/2]/Sqrt[2]
```

```
In[·]:= omega = 2 %
Out[·]:= Sqrt[2]*EllipticK[1/2]
```

On the other hand, ω is also equal to the following integral defined above:

```
In[·]:= 2 f[0, -1, 1]
Out[·]:= (2*Sqrt[Pi]*Gamma[5/4])/Gamma[3/4]
```

We can verify numerically that both expressions agree:

```
In[·]:= % - %% // N
Out[·]:= 0.
```

Symbolically,

```
In[·]:= FullSimplify[omega/2 == Integrate
           [1/Sqrt[1 - t^4], {t, 0, 1}]]
Out[·]:= True
```

The functions sin, sinh, sn are periodic; however, the periodicity of the hyperbolic sine is seen only in the complex plane:

```
In[·]:= Plot[{Sin[t], Sinh[t], JacobiSN
           [t, 1/3]}, {t, -10, 10}]
```

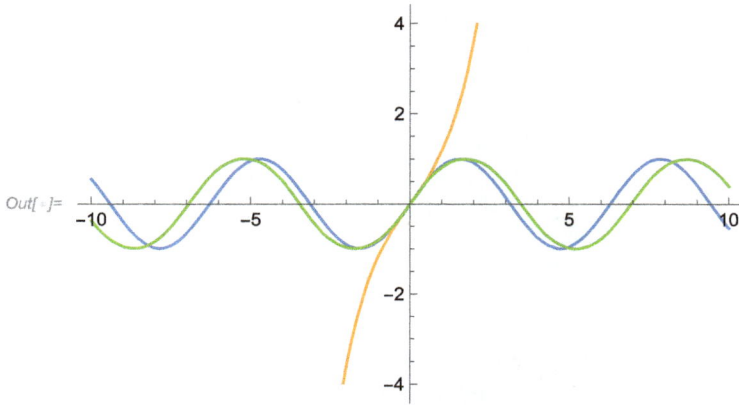

Figure 6.21

In[·]:= ComplexPlot[Sin[z], {z, -2 Pi - 6 I,
      2 Pi + 6 I}, PlotLegends -> Automatic]

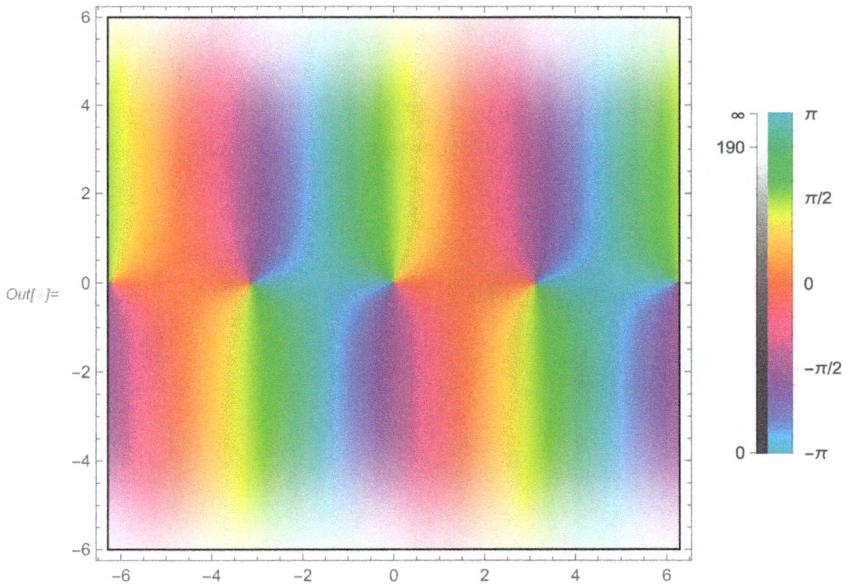

Figure 6.22

In[·]:= ComplexPlot[Sinh[z], {z, -2 Pi - 6 I,
      2 Pi + 6 I}, PlotLegends -> Automatic]

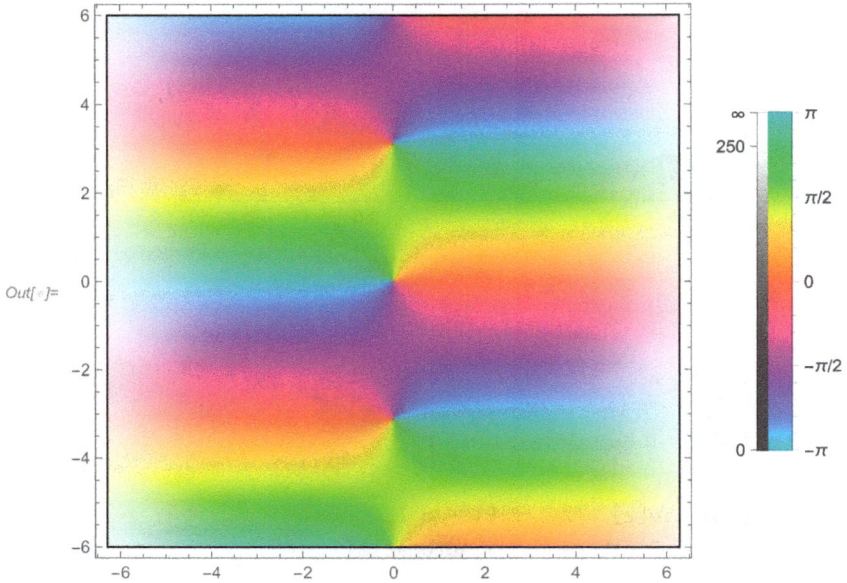

Figure 6.23

```
In[·]:= ComplexPlot[JacobiSN[z, 1/2], {z,
        -2 Pi - 6 I, 2 Pi + 6 I}, PlotLegends ->
        Automatic]
```

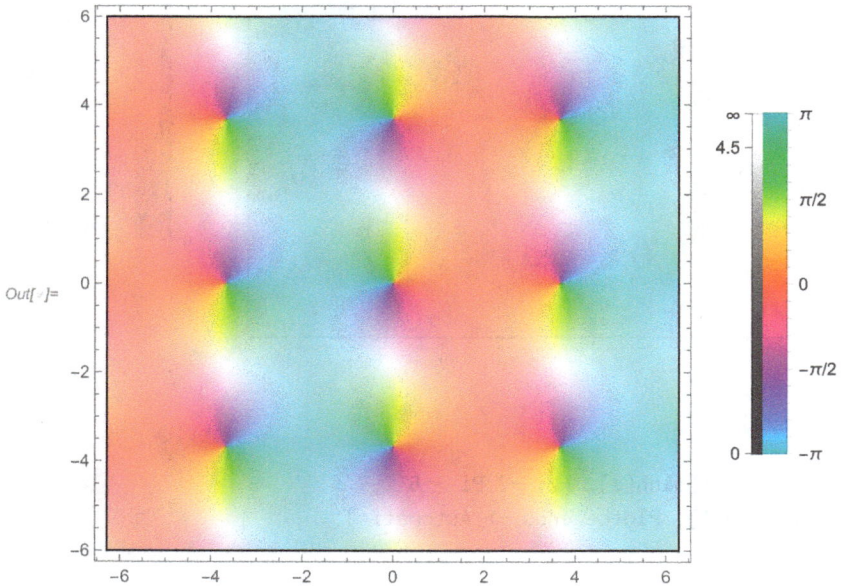

Figure 6.24

In[·]:= Sin[t + 2*Pi]
Out[·]:= Sin[t]

In[·]:= Sinh[t + 2*Pi*I]
Out[·]:= Sinh[t]

The functions sin and sinh are entire and they have only one period. Jacobi's sn function is, in fact, doubly periodic:

In[·]:= JacobiSN[t + 4*EllipticK[1/2], 1/2]
Out[·]:= JacobiSN[t, 1/2]

In[·]:= JacobiSN[t + 2*I*EllipticK[1/2], 1/2]
Out[·]:= JacobiSN[t, 1/2]

In[·]:= JacobiSN[t + 4*EllipticK[1/3], 1/3]
Out[·]:= JacobiSN[t, 1/3]

In[·]:= JacobiSN[t + 2*I*EllipticK[1 - 1/3], 1/3]
Out[·]:= JacobiSN[t, 1/3]

As it is proved in [13], the generalized sine function can be expressed in terms of the trigonometric, hyperbolic, and Jacobi's sn function. This can be achieved by a suitable substitution in the integral.

6.2 Algebraic addition theorems

The generalized sine function, $w = s(t)$, which is defined above as the inverse function to the integral $t = \int_0^w dz/\sqrt{1 + mz^2 + nz^4}$ with $m^2 - 4n \neq 0$, has the so-called algebraic addition formula (that is, one can express algebraically $s(y) = s(\alpha + \beta)$ in terms of $s(\alpha)$ and $s(\beta)$). Euler's method to find such a formula is to vary α and β in such a way that their sum y is constant. Then from the integral representation of $u = s(\alpha)$, $v = s(\beta)$, $w = s(\alpha + \beta)$ we have the equation $d(\alpha + \beta) = 0$, which is equivalent to the following ODE for the function $v(u)$:

In[·]:= equation = Derivative[1][v][u] ==
        -(Sqrt[1 + m*v[u]^2 + n*v[u]^4]/
        Sqrt[1 + m*u^2 + n*u^4]);

It is easy to show that the following function is the first integral:

In[·]:= F[u_, v_] := (u*Sqrt[1 + m*v^2 + n*v^4]
        + v*Sqrt[1 + m*u^2 + n*u^4])/
        (1 - n*u^2*v^2)

with initial conditions

In[·]:= {F[0, v] == v, F[u, 0] == u}
Out[·]:= {True, True}

We have

```
In[·]:= D[F[u, v[u]], u] /. (equation /.
          Equal -> Rule) // Together
Out[·]:= 0
```

We could have written the equation in the form with differentials *du* and *dv* and use the function Dt instead (so the variables *u* and *v* are treated independently):

```
In[·]:= Dt[F[u, v], Constants -> {m, n}] /.
          {Dt[u, Constants -> {m, n}] -> du,
           Dt[v, Constants -> {m, n}] -> dv};

In[·]:= Simplify[% /. dv -> (-du)*
          (Sqrt[1 + m*v^2 + n*v^4]/
           Sqrt[1 + m*u^2 + n*u^4])]
Out[·]:= 0
```

Let

```
In[·]:= F[u, v] /. {u -> s[a], v -> s[b]}
Out[·]:= (Sqrt[1 + m*s[a]^2 + n*s[a]^4]*s[b]
          + s[a]*Sqrt[1 + m*s[b]^2 + n*s[b]^4])
          /(1 - n*s[a]^2*s[b]^2)
```

and define

```
In[·]:= addForm[s_, a_, b_, m_, n_] :=
          s[a + b] - (Sqrt[1 + m*s[a]^2 +
          n*s[a]^4]*s[b] + s[a]*Sqrt[1 +
          m*s[b]^2 + n*s[b]^4])/
          (1 - n*s[a]^2*s[b]^2)
```

If we set this expression to zero, then this gives the algebraic addition formula for the generalized sine function. When we substitute particular values, we need to choose them in such a way that the algebraic addition formula is well-defined, that is the denominator is not zero. The following particular cases hold:

```
In[·]:= addForm[Sin, a, b, -1, 0]
Out[·]:= (-Sqrt[1 - Sin[a]^2])*Sin[b] -
          Sin[a]*Sqrt[1 - Sin[b]^2] +
          Sin[a + b]
```

which is the same as

```
In[·]:= Sin[a + b] // TrigExpand
Out[·]:= Cos[b] Sin[a] + Cos[a] Sin[b]
```

We can also substitute randomly chosen particular (complex) values for *a* and *b* to verify that addForm above is zero.

For the hyperbolic sine, we have the following addition formula:

In[·]:= `addForm[Sinh, a, b, 1, 0]`
Out[·]:= `(-Sqrt[1 + Sinh[a]^2])*Sinh[b] -`
`    Sinh[a]*Sqrt[1 + Sinh[b]^2] +`
`    Sinh[a + b]`

which is the same as

In[·]:= `Sinh[a + b] // TrigExpand`
Out[·]:= `Cosh[b] Sinh[a] + Cosh[a] Sinh[b]`

Substituting again random particular values, we see that `addForm` in this case is not zero. The branch of the square root can be chosen incorrectly as shown, for instance, by

In[·]:= `Chop[N[Simplify[-Sqrt[1 + Sinh[b]^2] -`
`    Cosh[b] /. {a -> 3/2, b -> 1 + 2*I}]]]`
Out[·]:= `0`

We remark that we have used the function `Chop` to replace the number close to zero by zero.

To find the addition formula for Jacobi's sn function, we need to substitute the following values:

In[·]:= `addForm[JacobiSN[#1, k^2] & ,`
`    a, b, -1 - k^2, k^2];`

The resulting expression is quite long so we omit it. Thus, we have seen that although **Mathematica**® knows a lot of formulas and identities for trigonometric, hyperbolic, and elliptic functions, in the case of some other formulas, not yet implemented, it may sometimes give incorrect results. One should not only rely completely on the symbolic or numerical output, one needs to consult other sources like the textbooks or various tables of formulas for verification.

6.3 The Weierstrass elliptic function

Many elliptic functions or related quantities like invariants, periods and functions like theta functions are already implemented in **Mathematica**® with related identities. However, some well-known identities are not known to **Mathematica**®. Below we give some examples taken from [2].

In[·]:= `Series[WeierstrassP[z, {g2, g3}],`
`    {z, 0, 3}]//Normal`
Out[·]:= `1/z^2 + (g2*z^2)/20`

In[·]:= `WeierstrassP[-z, {g2, g3}]`
Out[·]:= `WeierstrassP[z, {g2, g3}]`

It is well known that the function WeierstrassP is doubly periodic:

In[·]:= WeierstrassP[z + 2 Weierstrass
HalfPeriodW1[{g2, g3}], {g2, g3}]
Out[·]:= WeierstrassP[z, {g2, g3}]

In[·]:= WeierstrassP[z + 2 Weierstrass
HalfPeriodW2[{g2, g3}], {g2, g3}]
Out[·]:= WeierstrassP[z, {g2, g3}]

Defining e_1, e_2, and e_3 as values of the Weierstrass function at half-periods, for instance,

In[·]:= WeierstrassP[WeierstrassHalf
PeriodW1[{g2, g3}], {g2, g3}]
Out[·]:= WeierstrassE1[{g2, g3}]

the following relations should give zero:

In[·]:= WeierstrassE1[{g2, g3}] +
WeierstrassE2[{g2, g3}] +
WeierstrassE3[{g2, g3}] // Simplify
Out[·]:= WeierstrassE1[{g2, g3}] +
WeierstrassE2[{g2, g3}] +
WeierstrassE3[{g2, g3}]

The product of e_1, e_2, and e_3 should be equal to 1/4 WeierstrassInvariantG3[g2,g3], but Mathematica®'s version 12.3 does not automatically simplify this expression.

The absolute invariant J is not recognized if we input the definition directly (Mathematica® returns the input):

In[·]:= WeierstrassInvariantG2[{g2, g3}]^3/
(WeierstrassInvariantG2[{g2, g3}]^3 -
27*WeierstrassInvariantG3[{g2, g3}]^2);

However, there is a function KleinInvariantJ which is the function of the ratio of half-periods $\tau = \omega_2/\omega_1$ and which is suitable for both symbolic and numerical manipulation:

In[·]:= KleinInvariantJ[t + 1]
Out[·]:= KleinInvariantJ[t]

In[·]:= Series[KleinInvariantJ[x], {x, 0, 2}] // Normal
Out[·]:= (744 + 196884/E^((2*I*Pi)/x) + E^((2*I*Pi)/x))/1728

However,

In[·]:= KleinInvariantJ[-1/t]
Out[·]:= KleinInvariantJ[-(1/t)]

It would be nice if the relative invariants g_2 and g_3 were also defined as functions of the ratio of periods (or as functions of half-periods) to make it possible to study directly the action of modular substitutions.

We can plot the invariants in the following way:

In[·]:= `Plot[WeierstrassInvariantG3[{1, I t}],`
        `{t, 1, 4}]`

Out[ ]=

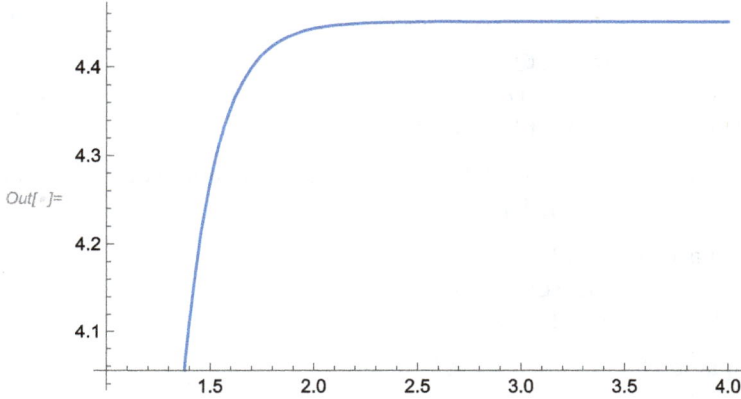

Figure 6.25

or

In[·]:= `Plot[Im[WeierstrassInvariants[{w1,`
        `1 + I}]], {w1, -3, 3}]`

Out[ ]=

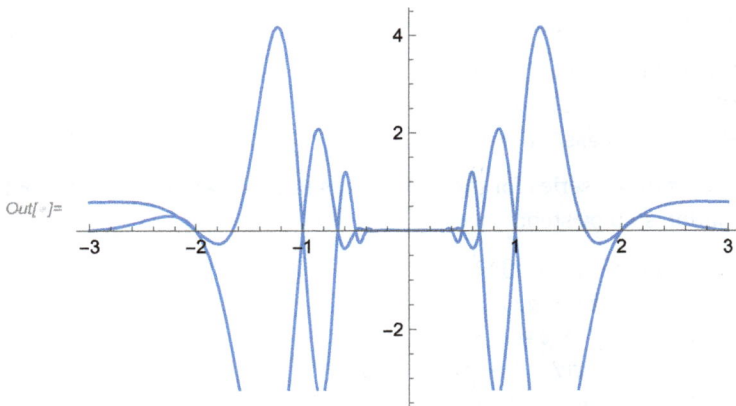

Figure 6.26

It is known that for any finite c equation $J(\tau) = c$ has one and only one root in some region D [2, Par. 10]. Mathematica® can also solve the following equation:

In[·]:= Solve[KleinInvariantJ[t] == 2, t]

⋯ Solve: Inverse functions are being used by Solve, so some solutions may not be found; use Reduce for complete solution information.

Out[ ]= $\left\{\left\{t \rightarrow \boxed{⌕\ \text{21.0...}\ +\ \text{1.25...}\ \text{i}}\right\}\right\}$

Figure 6.27

However,

In[·]:= Reduce[KleinInvariantJ[t] == 2, t]

⋯ Reduce: This system cannot be solved with the methods available to Reduce.

Out[·]:= Reduce[KleinInvariantJ[t] == 2, t]

Note that in addition to KleinInvariantJ there are several other built-in functions like ModularLambda or DedekindEta which depend on τ:

In[·]:= D[ModularLambda[t], t]
Out[·]:= (4*I*EllipticK[ModularLambda[t]]^2*
      (1 - ModularLambda[t])*ModularLambda[t])/Pi

The differential equation for the Weierstrass function is

In[·]:= DSolve[Derivative[1][y][z]^2 ==
      4*y[z]^3 - g2*y[z] - g3, y[z], z]
Out[·]:= {{y[z] -> WeierstrassP[z - C[1], {g2, g3}]},
      {y[z] -> WeierstrassP[z + C[1], {g2, g3}]}}

Note that the following second-order nonlinear differential equation can also be solved in terms of the Weierstrass elliptic function (just differentiate the previous equation and cancel y' on both sides):

In[·]:= DSolve[Derivative[2][y][z] ==
      6*y[z]^2 - g2/2, y[z], z]
Out[·]:= {{y[z] -> WeierstrassP[z + C[1], {g2, C[2]}]}}

We can find coefficients of the series for the Weierstrass function above directly from the differential equation step by step:

In[·]:= Series[Derivative[1][y][z]^2 -
      (4*y[z]^3 - g2*y[z] - g3) /. {y[z] ->
      a1/z^2 + a2/z + a3 + a4*z + a5*z^2 +
      a6*z^3, D[y[z] -> a1/z^2 + a2/z + a3 +
       a4*z + a5*z^2 + a6*z^3, z]}, {z, 0, 2}]
      /. {a1 -> 1, a2 -> 0, a3 -> 0, a4 -> 0,
      a5 -> g2/20} // Normal
Out[·]:= g3 - (24*a6)/z + (3*g2^2*z^2)/100

To determine the coefficient a_6, we need to take more coefficients in the expansion.

The Weierstrass elliptic function uniformizes the following elliptic curve:

```
In[·]:= Manipulate[ContourPlot[y^2 == x^3 +
       a*x + b, {x, -10, 10}, {y, -10, 10}],
       {a, -10, 10}, {b, -10, 10}]
```

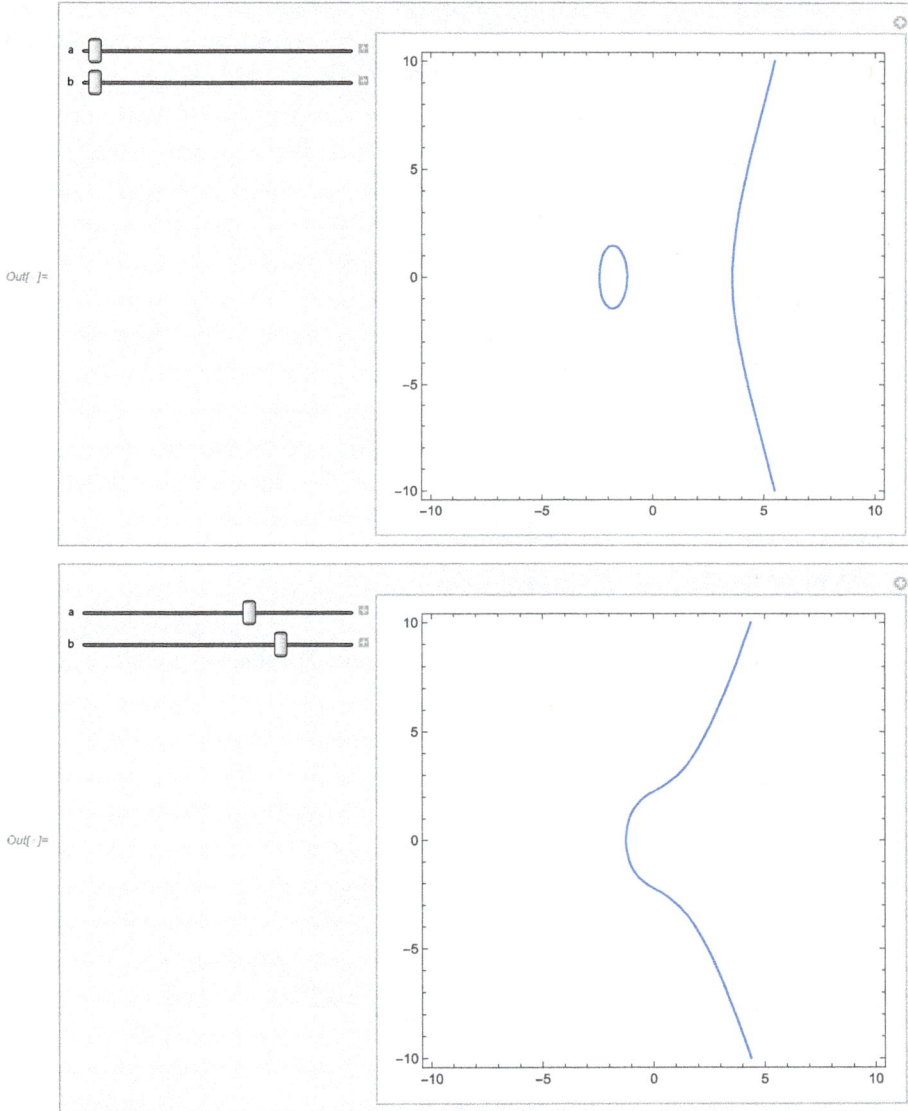

Out[ ]=

Figure 6.28

Mathematica® can sum over positive and negative integers:

In[·]:= Sum[1/n^2, {n, 1, Infinity}]
Out[·]:= Pi^2/6

In[·]:= Sum[1/n^2, {n, -Infinity, -1}]
Out[·]:= Pi^2/6

However, when we input the following expressions, they are not recognized as an elliptic function or its invariants (or, more precisely, using the terminology of [2] relative invariants in comparison with the absolute invariant *J*):

In[·]:= 1/z^2 + Sum[1/(z - 2*m - 2*m*I)^2 -
        1/(2*m + 2*m*I)^2, {m, 1, Infinity},
        {n, 1, Infinity}] + Sum[1/(z - 2*m
        - 2*m*I)^2 - 1/(2*m + 2*m*I)^2,
        {m, -Infinity, -1}, {n, -Infinity, -1}]

In[·]:= Sum[1/(2*n + 2*m*I)^4, {m, 1,
        Infinity}, {n, 1, Infinity}] +
        Sum[1/(2*n + 2*m*I)^4, {m,
        -Infinity, -1}, {n, -Infinity, -1}]

However, sometimes Mathematica® can calculate certain values explicitly:

In[·]:= WeierstrassInvariants[{1, I}]
Out[·]:= {Gamma[1/4]^8/(256*Pi^2), 0}

In[·]:= WeierstrassInvariants[{1, 2 I}]
Out[·]:= WeierstrassInvariants[{1, 2 I}]

Theta functions are implemented in Mathematica®, but still some expressions are not automatically simplified.

In[·]:= Plot[EllipticTheta[3, x, 1/2], {x, -2, 2}]

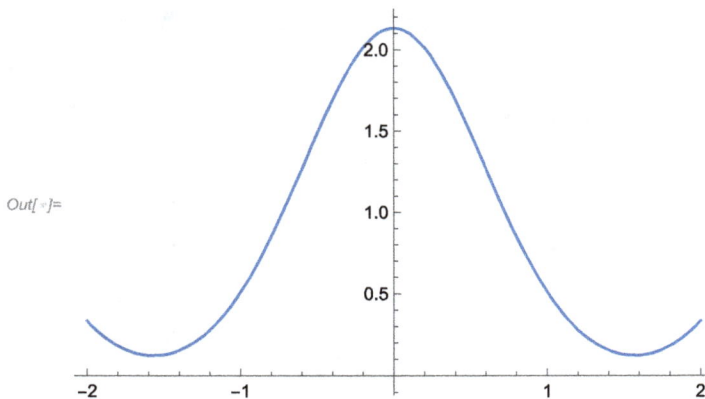

Out[·]=

Figure 6.29

```
In[·]:= Series[EllipticTheta[3, x, t], {t, 0, 4}]
          // Normal
Out[·]:= 1 + 2*t*Cos[2*x] + 2*t^4*Cos[4*x]
```

```
In[·]:= EllipticTheta[3, x + t, t] // FullSimplify
Out[·]:= EllipticTheta[3, t + x, t]
```

The new functions like FunctionMeromorphic and FunctionSingularities that appeared in version 12.2 of Mathematica® might need some more work when dealing with special functions. They work quite well with elementary functions. For instance, it is not known in the current version we use (Mathematica® 12.3) whether the Weierstrass function is meromorphic or not for general values of parameters (maybe some conditional expressions in the output could be returned). We hope that those functions will be updated and extended in newer versions to be consistent with the theory in standard textbooks.

```
In[·]:= FunctionMeromorphic[1/z, z]
Out[·]:= True
```

```
In[·]:= FunctionMeromorphic[WeierstrassP
          [z, {g2, g3}], z]
Out[·]:= FunctionMeromorphic[WeierstrassP
          [z, {g2, g3}], z]
```

```
In[·]:= FunctionMeromorphic[WeierstrassP
          [z, {1/2, 3}], z]
Out[·]:= True
```

```
In[·]:= FunctionMeromorphic[EllipticTheta
          [3, x, 1/2], x]
Out[·]:= True
```

```
In[·]:= FunctionMeromorphic[EllipticTheta
          [3, x, t], x]
Out[·]:= ConditionalExpression[True,
          Im[t]^2 + Re[t]^2 < 1]
```

```
In[·]:= FunctionSingularities[1/z, z]
Out[·]:= z == 0
```

```
In[·]:= FunctionSingularities[E^z, z,
          Complexes]
Out[·]:= False
```

In[·]:= FunctionSingularities[WeierstrassP
    [z, {g2, g3}], z, Reals]

Out[·]= Sin[(π z Im[WeierstrassHalfPeriodW3[{g2, g3}]]) /
    (2 (Im[WeierstrassHalfPeriodW3[{g2, g3}]] Re[WeierstrassHalfPeriodW1[{g2, g3}]] –
        Im[WeierstrassHalfPeriodW1[{g2, g3}]] Re[WeierstrassHalfPeriodW3[{g2, g3}]])))] == 0

Figure 6.30

When we change the domain in the expression above to Complexes, the answer is even longer.

In[·]:= FunctionSingularities[EllipticTheta
    [3, x, t], x, Complexes]
Out[·]:= Im[t]^2 + Re[t]^2 >= 1

In[·]:= FunctionSingularities[EllipticTheta
    [3, x, t], x, Reals]
Out[·]:= t >= 1 || t <= -1

In[·]:= FunctionSingularities[EllipticTheta
    [3, x, 1/2], x, Complexes]
Out[·]:= False

Elliptic functions often appear in coefficients of differential equations. For instance, in the elliptic form of the Lamé equation and, more generally, in the elliptic form of the Heun equation. The Lamé equation is in the following form, where n is a nonnegative integer and l is an arbitrary constant, however, Mathematica® does not recognize it, returning the differential equation unsolved:

In[·]:= DSolve[Derivative[2][y][z] ==
    (n*(n + 1)*WeierstrassP[z, {1/2, 10}]
    + l)*y[z] /. n -> 1, y[z], z]

However, as we can see below, it is implemented slightly differently in Mathematica® [10].

In[·]:= DSolve[Derivative[2][y][z] +
    (h - v*(v + 1)*m*JacobiSN[z, m]^2)*
    y[z] == 0, y[z], z]

Out[·]= $\left\{\left\{y[z] \to c_1\, \text{HeunG}\left[\frac{1}{m}, -\frac{h}{4\,m}, -\frac{v}{2}, \frac{1+v}{2}, \frac{1}{2}, \frac{1}{2}, \text{JacobiSN}[z, m]^2\right] + \right.\right.$
$\left.\left. c_2\, \text{HeunG}\left[\frac{1}{m}, \frac{1}{4} + \frac{1-h}{4\,m}, \frac{1-v}{2}, 1+\frac{v}{2}, \frac{3}{2}, \frac{1}{2}, \text{JacobiSN}[z, m]^2\right]\sqrt{\text{JacobiSN}[z, m]^2}\right\}\right\}$

Figure 6.31

```
In[·]:= DSolve[Derivative[2][y][z] +
        (LameEigenvalueA[v, j, m] - v*(v + 1)*m*
        JacobiSN[z, m]^2)*y[z] == 0, y[z], z]
```

$$Out[\ ]= \left\{\left\{y[z] \rightarrow c_1 \, \text{LameC}[v, j, z, m] + c_2 \, \text{LameC}[v, j, z, m] \int_1^z \frac{1}{\text{LameC}[v, j, K[1], m]^2} \, dK[1]\right\}\right\}$$

Figure 6.32

6.4 Example: Addition formulas for the Weierstrass elliptic function

Let $\wp(u)$ be the Weierstrass elliptic function satisfying

```
In[·]:= eq = Derivative[1][p][u]^2 ==
        4*p[u]^3 - g2*p[u] - g3;
```

Let us first prove the following formula:

```
In[·]:= addform = p[z1 + z2] == (1/4)*
        ((Derivative[1][p][z1] - Derivative[1]
        [p][z2])/(p[z1] - p[z2]))^2 -
        p[z1] - p[z2];
```

We will use the differential equation several times for variables z_1 and z_2. Moreover, we shall replace the terms $(\wp')^k$, where $k \geq 3$, again by using the differential equation.

```
In[·]:= subst1[z_] := (eq /. u -> z /.
        Equal -> Rule)
```

```
In[·]:= subst2[z_] := Expand[Table[subst1[z]
        [[1]]*Derivative[1][p][z]^i ->
        subst1[z][[2]]*Derivative[1][p][z]^i,
        {i, 0, 4}]]
```

```
In[·]:= (eq /. u -> z1 + z2) /. {addform /.
        Equal -> Rule, D[addform /. Equal ->
        Rule, z1]} // Simplify;
```

We also need to replace the second-order derivative of the function $\wp$:

```
In[·]:= Together[Simplify[% /. Solve[
        Factor[D[eq /. u -> z1, z1]],
        Derivative[2][p][z1]][[1]][[1]]
        //. subst1[z1]]]];
```

Finally, substituting everything we complete the proof:

In[·]:= % //. subst2[z1] //. subst2[z2]
        // Simplify // Factor
Out[·]:= True

Another formula that we will verify numerically is given by

In[·]:= Chop[N[Det[{{1, WeierstrassP[u1,
        {g2, g3}], D[WeierstrassP[u1, {g2,
        g3}], u1]}, {1, WeierstrassP[u2,
        {g2, g3}], D[WeierstrassP[u2, {g2,
        g3}], u2]}, {1, WeierstrassP[-u1
        - u2, {g2, g3}], D[WeierstrassP[u3,
        {g2, g3}], u3] /. u3 -> -u1 - u2}}]
        /. {g2 -> RandomReal[], g3 ->
        RandomReal[], u1 -> RandomReal[],
        u2 -> RandomReal[]}]]
Out[·]:= 0

7 Elements of complex analysis

Although our book series is devoted to real analysis, we often refer to topics from complex analysis. In this chapter we would like to give a few more examples and dynamic illustrations to demonstrate how Mathematica® can be used with functions of complex variables. We would like to remark that a very detailed account of the theory with Mathematica® examples is in the book [16]. Other references used are [3, 14].

7.1 Complex numbers and functions

We can form complex numbers from a pair of real numbers using the function `Complex`:

In[·]:= `Complex @@ {1, 1}`
Out[·]:= `1 + I`

In[·]:= `Complex @@@ {{1, 1}, {2, 2}}`
Out[·]:= `{1 + I, 2 + 2*I}`

The inverse operation is

In[·]:= `ReIm[%]`
Out[·]:= `{{1, 1}, {2, 2}}`

A random complex number with real and imaginary part in the interval $[0, 1]$ is given by

In[·]:= `RandomComplex[]`
Out[·]:= `0.483331 + 0.49445 I`

and a random number from a rectangle (specifying the lower left and the upper right corners) is given by

In[·]:= `RandomComplex[{-100 - 100*I, 100 + 100*I}]`
Out[·]:= `41.34458277362211 - 27.696852687579735*I`

The complex infinity often appears in outputs of expressions like

In[·]:= `1/0`
　　　`··· Power: Infinite expression 1/0 encountered.`
Out[·]:= `ComplexInfinity`

This represents a quantity with infinite magnitude, but undetermined complex argument.

Mathematica® can compute absolute values and arguments:

In[·]:= `AbsArg[3 + 4*I]`
Out[·]:= `{5, ArcTan[4/3]}`

The function `Conjugate` gives the conjugate complex number:

https://doi.org/10.1515/9783110774641-007

In[·]:= `Conjugate[5 + 12*I]`
Out[·]:= `5 - 12*I`

Complex numbers often appear when solving algebraic equations:

In[·]:= `Solve[x^2 - 2*x + 3 == 0, x]`
Out[·]:= `{{x -> 1 - I*Sqrt[2]},`
`        {x -> 1 + I*Sqrt[2]}}`

In[·]:= `ToRules[Reduce[x^2 - 2*x + 3 == 0, x]]`
Out[·]:= `Sequence[{x -> 1 - I*Sqrt[2]},`
`        {x -> 1 + I*Sqrt[2]}]`

In[·]:= `{ToRules[Reduce[x^2 - 2*x + 3 == 0, x]]}`
Out[·]:= `{{x -> 1 - I*Sqrt[2]},`
`        {x -> 1 + I*Sqrt[2]}}`

There exists a couple of useful built-in functions that work with complex symbolic expressions, for instance,

In[·]:= `ComplexExpand[Sin[x + I*y]]`
Out[·]:= `Cosh[y]*Sin[x] + I*Cos[x]*Sinh[y]`

In[·]:= `TrigReduce[%]`
Out[·]:= `Sin[x + I*y]`

In[·]:= `ExpToTrig[Exp[I*x]]`
Out[·]:= `Cos[x] + I*Sin[x]`

In[·]:= `TrigToExp[Cos[x]]`
Out[·]:= `1/(E^(I*x)*2) + E^(I*x)/2`

Most built-in functions can also take complex numbers as arguments, for instance,

In[·]:= `Limit[Sin[z - I]/(z - I), z -> I]`
Out[·]:= `1`

In[·]:= `D[Sin[I*z], z]`
Out[·]:= `I*Cosh[z]`

In[·]:= `Normal[Series[z^2, {z, I, 2}]]`
Out[·]:= `-1 + 2*I*(-I + z) + (-I + z)^2`

In[·]:= `Residue[1/(z - I), {z, I}]`
Out[·]:= `1`

The integration along lines joining complex numbers can be done as follows:

In[·]:= `Integrate[z, {z, I, 1 + I, 2 + 3*I}]`
Out[·]:= `-2 + 6*I`

7.2 Example 1: Möbius transformation

Let us show that every Möbius transformation is a composition of several transformations consisting only of similarities of the extended complex plane and complex inversion.

```
In[·]:= Composition[Function[z, z + a/c],
        Function[z, -(a d - b c) z/c^2],
        Function[z, 1/z], Function[z,
        z + d/c]][z] // Together
Out[·]:= (b + a*z)/(d + c*z)
```

We need to be careful about the order of composition of functions:

```
In[·]:= Composition[Function[y, y + 2],
        Function[y, 1/y]][z]
Out[·]:= 2 + 1/z
```

```
In[·]:= Composition[Function[y, 1/y],
        Function[y, y + 2]][z]
Out[·]:= 1/(2 + z)
```

See also [30].

7.3 Example 2: The conformal mapping $w = z^2$

Let $f : \mathbb{C} \to \mathbb{C}$ be the complex function given by $f : z \to z^2$. We shall use the following notation: $w = f(z) = z^2$, $z = x + iy = re^{i\phi}$, $w = u + iv = \rho e^{i\psi}$, $\rho = r^2$, $\psi = 2\phi$, and $i^2 = -1$.
(i) Find the images of the half-line $x = x_0 \geq 0$, $y > 0$ and of the line $y = y_0 > 0$, $x \in \mathbb{R}$.
(ii) Calculate explicitly an angle between the resulting two curves at the point $f(z_0)$, where $z_0 = x_0 + iy_0$, $x_0 > 0$, $y_0 > 0$, is the point of intersection of the given two lines.

The problem can easily be reformulated in terms of differential geometry of plane curves. First, let us visualize the original (half-)lines. Take $z = x + iy$. The red half-line is given by $x = x_0 \geq 0$, $y > 0$. By changing x_0, we make it move in the first quadrant, and the yellow point indicates that we are not considering the point with coordinates $(x_0, 0)$. The green line is the line given by $y = y_0 > 0$, $x \in \mathbb{R}$. By changing y_0, we make the given line move in the upper half-plane. Clearly, the lines intersect at a right angle at some point (x_0, y_0) with $x_0 > 0$, $y_0 > 0$.

In[·]:= Manipulate[Show[ParametricPlot
       [{x0, t}, {t, 0.1, 10}, PlotStyle ->
       Red, AxesLabel -> {x, y}, PlotRange
       -> {{-10, 10}, {-1, 10}}],
       ParametricPlot[{t, y0}, {t, -10, 10},
       PlotStyle -> Green, PlotRange ->
       {{-10, 10}, {-10, 10}}], Graphics
       [{PointSize[Large], Yellow,
       Point[{x0, 0}]}]], {x0, 0, 10},
       {y0, 0.2, 10}, SaveDefinitions -> True]

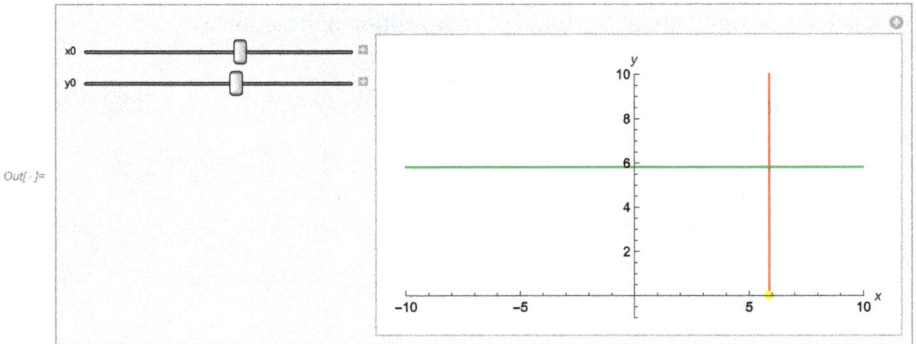

Figure 7.1

We will use complex numbers to visualize points in $\mathbb{R}^2$ and develop some intuition for our problem. We shall plot several points together with their images under the map f. First, let us take the point with coordinates $(1, 1)$ (or $z = 1 + i$).

In[·]:= pointz = {1, 1}
Out[·]:= {1, 1}

Its image is the point $(0, 2)$ (or $w = 2i$).

In[·]:= pointw = ({Re[#1^2], Im[#1^2]} &)
       [pointz . {1, I}]
Out[·]:= {0, 2}

Alternatively, we can use the function ReIm:

In[·]:= (ReIm[#1^2] &)[pointz . {1, I}]
Out[·]:= {0, 2}

Let us visualize both points.

In[·]:= points[pointz_, pointw_, plotrange_] :=
    Show[{ListLinePlot[{{0, 0}, pointz},
    PlotStyle -> Pink, PlotRange ->
    plotrange, AspectRatio -> Automatic],
    ListLinePlot[{{0, 0}, pointw},
    PlotStyle -> Blue, PlotRange ->
    plotrange], Graphics[{{PointSize[Large],
    Pink, Point[pointz], Text[z,
    pointz + 1/15]}, {Blue, PointSize
    [Large], Point[pointw], Text[w ==
    z^2, pointw + 1/10]}}, Axes
    -> True, PlotRange -> plotrange]}]

In[·]:= points[pointz, pointw, {{-0.5, 1.5},
    {0, 2.5}}]

Out[·]=

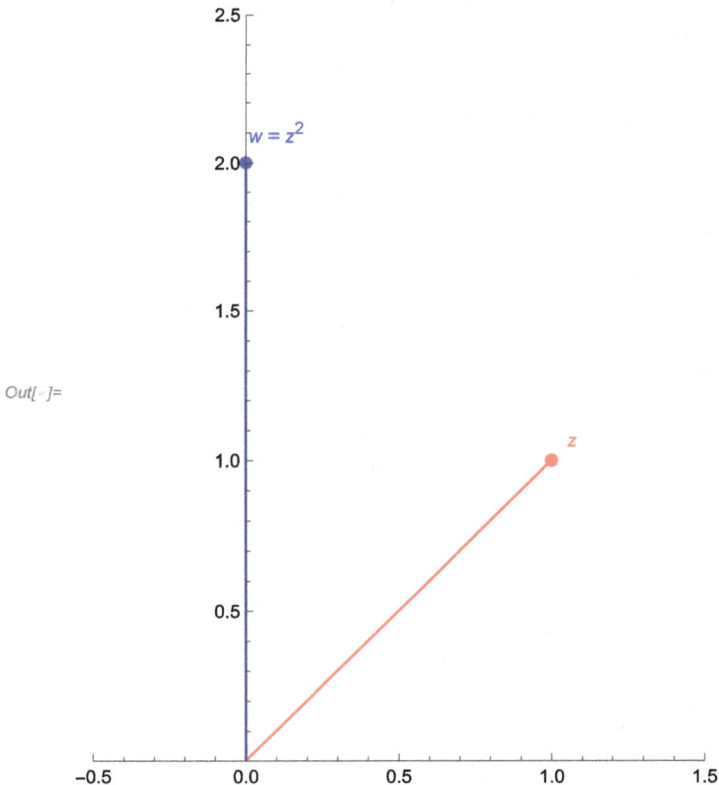

Figure 7.2

Notice the change of the modulus and of the argument of both complex numbers: $z = x + iy = re^{i\phi}$, $w = u + iv = \rho e^{i\psi}$, $\rho = r^2$, $\psi = 2\phi$, and $i^2 = -1$.

In[·]:= pointz . {1, I}
Out[·]:= 1 + I

In[·]:= {Abs[%], Arg[%]}
Out[·]:= {Sqrt[2], Pi/4}

Alternatively, we can use the function AbsArg:

In[·]:= AbsArg[%%]
Out[·]:= {Sqrt[2], Pi/4}

In[·]:= {Abs[pointw . {1, I}],
        Arg[pointw . {1, I}]}
Out[·]:= {2, Pi/2}

Let us draw several more points which lie on the line $x = 1$.

In[·]:= pointz1 = {1, 1/4}
Out[·]:= {1, 1/4}

In[·]:= pointw1 = ({Re[#1^2], Im[#1^2]} &)
        [pointz1 . {1, I}]
Out[·]:= {15/16, 1/2}

In[·]:= points[pointz1, pointw1,
        {{-0.5, 1.5}, {0, 1}}]

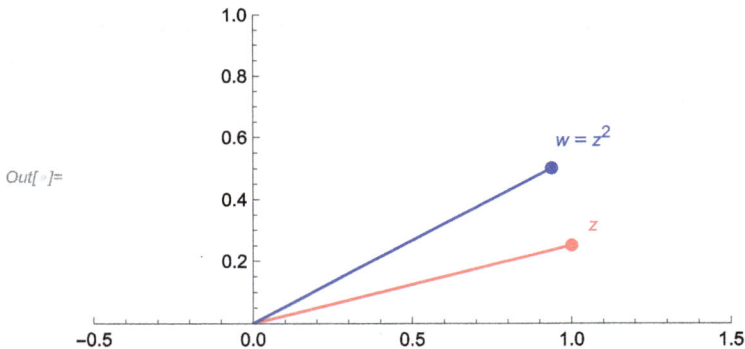

Figure 7.3

In[·]:= pointz2 = {1, 1.5}
Out[·]:= {1, 1.5}

In[·]:= pointw2 = ({Re[#1^2], Im[#1^2]} &)
        [pointz2 . {1, I}]
Out[·]:= {-1.25, 3.}

In[·]:= `points[pointz2, pointw2, {{-2, 1.5},`
        `{0, 3.2}}]`

Out[·]=

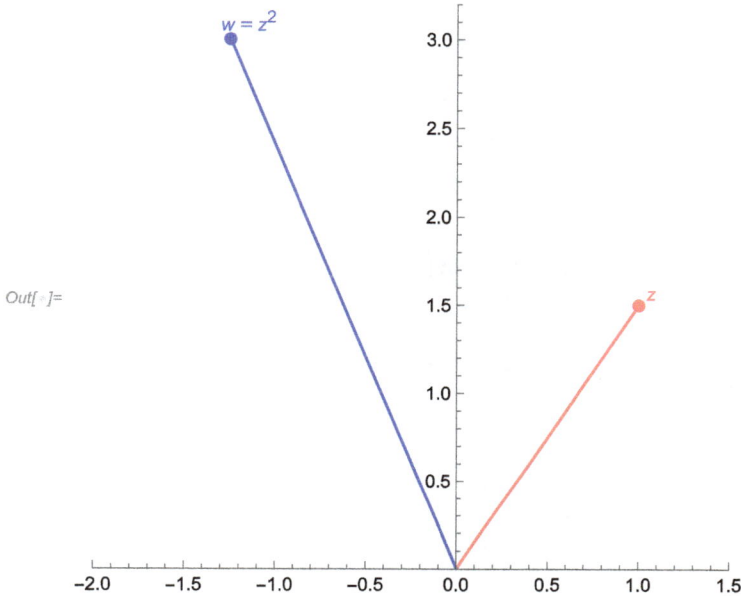

Figure 7.4

We see from our experiments that the images of the points on the half-line $x = x_0 > 0$, $y > 0$ cannot lie on a straight line. The points with the imaginary part less than 1 might be quite close to the points to the original points on the half-line, whereas the points with the imaginary part greater than 1 might be located quite far.

Next let us study how the points on the upper half of the imaginary axis change under the transformation f. The real and imaginary parts of $w = f(z)$ are

In[·]:= `Expand[(x + I*y)^2]`
Out[·]:= `x^2 + 2*I*x*y - y^2`

In[·]:= `fuv = {u, v} == Refine[{Re[%],`
          `Im[%]}, Element[x || y, Reals]]`
Out[·]:= `{u, v} == {x^2 - y^2, 2*x*y}`

Alternatively, we can use

In[·]:= `ComplexExpand[ReIm[(x + I y)^2]]`
Out[·]:= `{x^2 - y^2, 2*x*y}`

For the half-line $x = 0$, $y > 0$, we have

In[·]:= `fuv /. {x -> 0}`
Out[·]:= `{u, v} == {-y^2, 0}`

So the image of the half-line $x = 0$, $y > 0$ will be the left half of the horizontal axis (without the origin), and we can visualize this in the following way. Notice how the corresponding points on these lines move.

```
In[·]:= Manipulate[Show[ParametricPlot[
        {0, t}, {t, 0.1, 10}, PlotStyle ->
        Pink, PlotRange -> {{-10, 10},
        {-1, 10}}], Graphics[{PointSize[
        Large], Yellow, Point[{0, 0}]}],
        Graphics[{PointSize[Large], Pink,
        Point[{0, y}]}], ParametricPlot[
        {-t^2, 0}, {t, 0.1, 10}, PlotStyle ->
        Blue, PlotRange -> {{-10, 10}, {-1,
        10}}], Graphics[{PointSize[Large],
        Blue, Point[{-y^2, 0}]}]], {y, 0.4,
        10}, SaveDefinitions -> True]
```

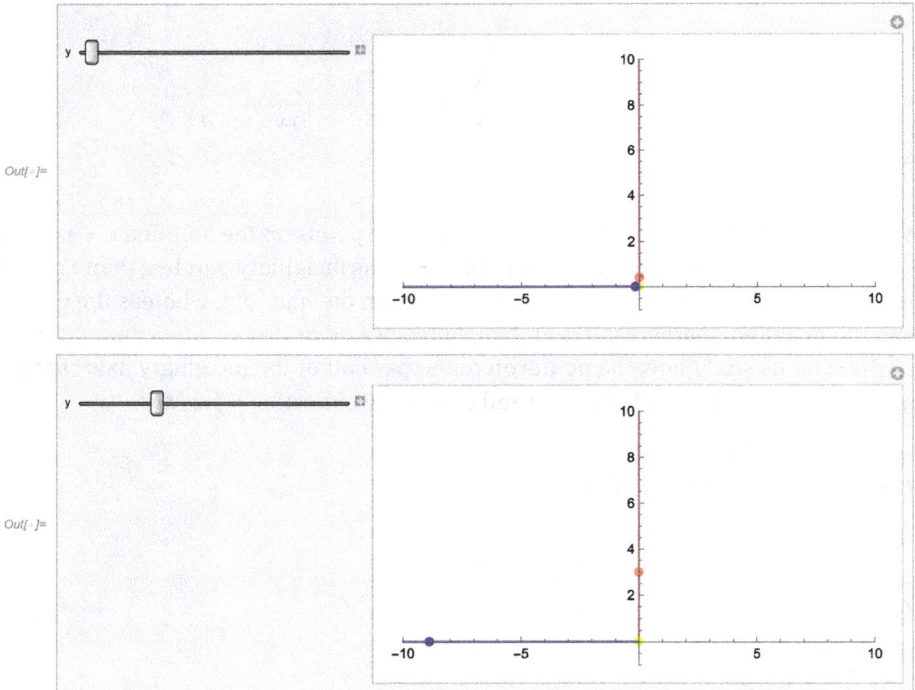

Figure 7.5

Now let us find the image of the half-line $x = x_0 > 0, y > 0$.

```
In[·]:= Show[ParametricPlot[{1, t}, {t,
        0.01, 2}, PlotStyle -> Pink,
        AxesLabel -> {x, y}, PlotRange ->
        {{-0.5, 3/2}, {-2^(-1), 1/2}}],
        Graphics[{PointSize[Large],
        Yellow, Point[{1, 0}]}]]
```

Out[ ]=

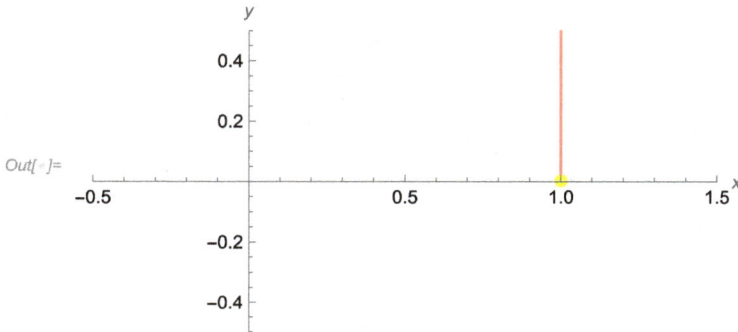

Figure 7.6

```
In[·]:= fuv /. x -> x0
Out[·]:= {u, v} == {x0^2 - y^2, 2*x0*y}
```

Eliminating y from these equations gives an equation for a horizontal (sideways) parabola (oriented to the left-hand side):

```
In[·]:= eq1 = Eliminate[%, y] // Simplify
Out[·]:= v^2 + 4*u*x0^2 == 4*x0^4
```

Notice, however, that $x_0 > 0$ and $y > 0$, so the v coordinate should be positive.

```
In[·]:= fuv /. {x -> x0}
Out[·]:= {u, v} == {x0^2 - y^2, 2*x0*y}
```

In fact, we should have only the upper half of this parabola (without the point $(x_0^2, 0)$) as the image of the half-line $x = x_0 > 0, y > 0$. For $x_0 = 1$, we have

```
In[·]:= eq1 /. {x0 -> 1}
Out[·]:= 4*u + v^2 == 4
```

```
In[·]:= ContourPlot[Evaluate[%],
        {u, -1, 1}, {v, 0, 3}]
```

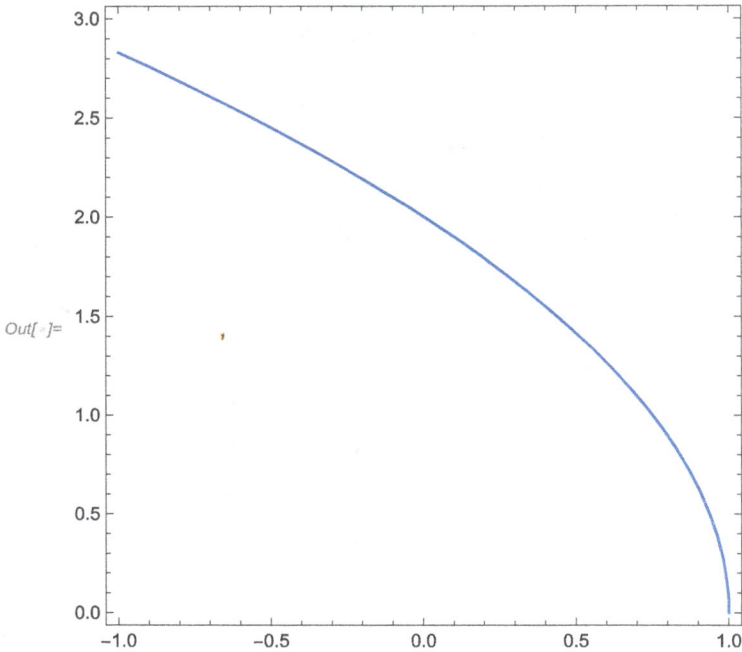

Figure 7.7

Now let us visualize the original and the image curves and the points on them.

```
In[·]:= Manipulate[Show[ParametricPlot[
        {x0, t}, {t, 0.1, 8}, PlotStyle ->
        Pink, PlotRange -> {{-10, 5},
        {-1, 8}}], Graphics[{PointSize[
        Large], Yellow, Point[{x0, 0}]}],
        Graphics[{PointSize[Large], Yellow,
        Point[{x0^2, 0}]}], Graphics[
        {PointSize[Large], Pink, Point[
        {x0, y}]}], Graphics[{PointSize[
        Large], Blue, Point[{x0^2 - y^2,
        2*x0*y}]}], ContourPlot[v^2 +
        4*u*x0^2 == 4*x0^4, {u, -10, 5},
        {v, 0, 8}]], {y, 0.3, 10}, {x0, 0.1,
        5, 0.4}, SaveDefinitions -> True]
```

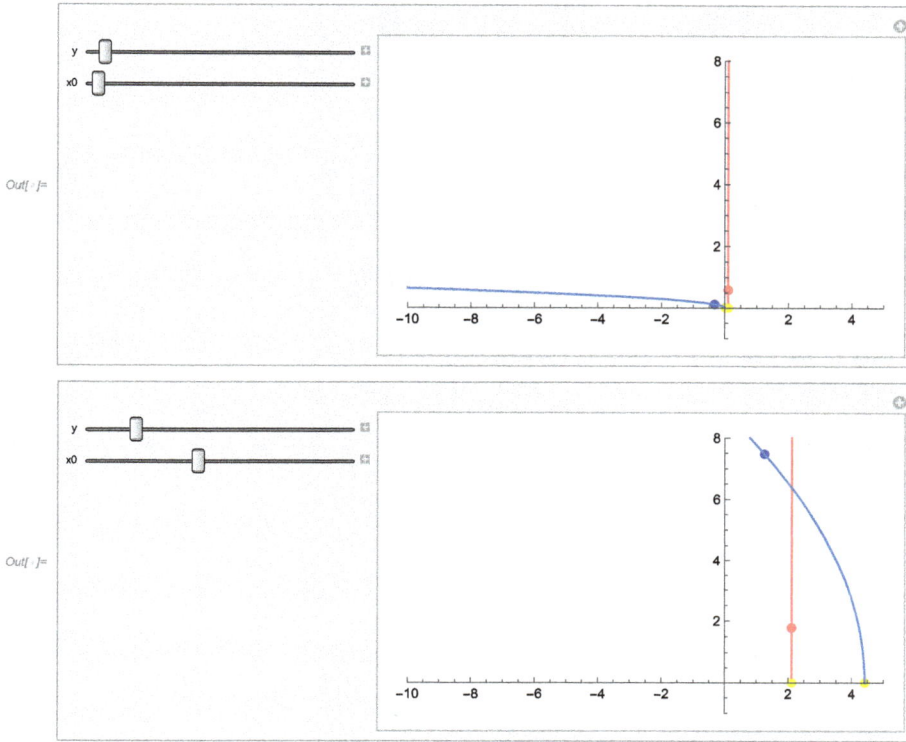

Figure 7.8

Let us now find the image of the second horizontal line $y = y_0 > 0$, $x \in \mathbb{R}$. We shall also visualize the results for $y_0 = 1$ and for the general case.

```
In[·]:= fuv /. y -> y0
Out[·]:= {u, v} == {x^2 - y0^2, 2*x*y0}
```

```
In[·]:= eq2 = Eliminate[%, x] // Simplify
Out[·]:= v^2 == 4*y0^2*(u + y0^2)
```

```
In[·]:= v^2 == 4*y0^2*(u + y0^2) /. y0 -> 1
Out[·]:= v^2 == 4*(1 + u)
```

In[·]:= ContourPlot[Evaluate[%],
　　　　{u, -1, 1}, {v, -4, 5}]

Out[ ]=

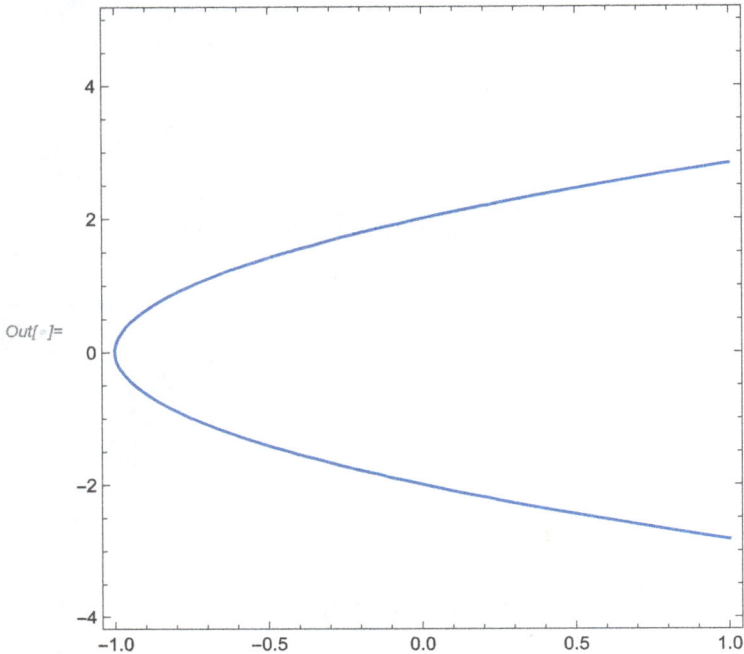

Figure 7.9

In[·]:= Manipulate[Show[ParametricPlot[
　　　　{t, 1}, {t, -10, 10}, PlotStyle
　　　　-> Pink, PlotRange -> {{-10, 10},
　　　　{-10, 10}}], Graphics[{PointSize[
　　　　Large], Pink, Point[{y, 1}]}],
　　　　Graphics[{PointSize[Large], Blue,
　　　　Point[{y^2 - 1^2, 2*1*y}]}],
　　　　ContourPlot[v^2 == 4*(1 + u),
　　　　{u, -10, 10}, {v, -10, 10}]], {y,
　　　　-10, 10, 1}, SaveDefinitions -> True]

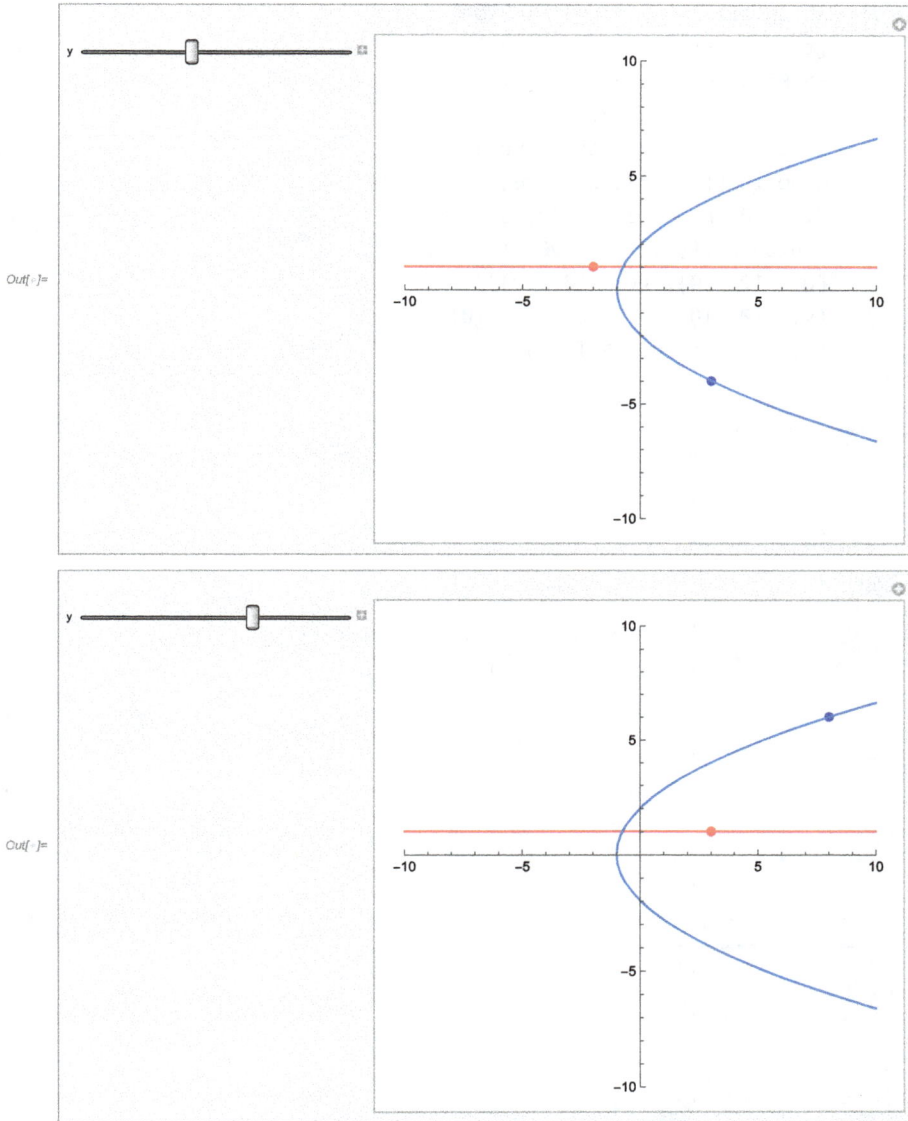

$Out[\circ]=$

$Out[\circ]=$

Figure 7.10

```
In[·]:= Manipulate[Show[ParametricPlot[
        {t, y0}, {t, -10, 10}, PlotStyle
        -> Pink, PlotRange -> {{-10, 10},
        {-10, 10}}], Graphics[{PointSize[
        Large], Pink, Point[{y, y0}]}],
        Graphics[{PointSize[Large],
        Blue, Point[{y^2 - y0^2, 2*y*y0}]}],
        ContourPlot[v^2 == 4*y0^2*(u + y0^2),
        {u, -10, 10}, {v, -10, 10}]],
        {y, -10, 10, 1}, {y0, 0.3, 10},
        SaveDefinitions -> True]
```

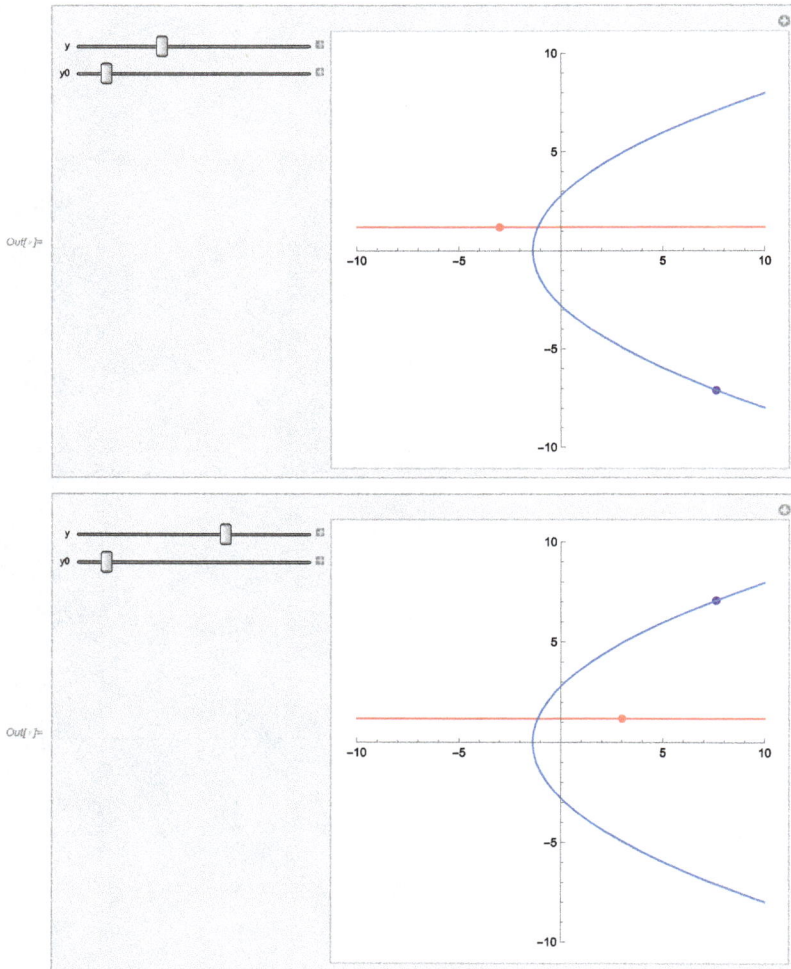

Figure 7.11

So the original (half-)lines (pink color) and the resulting (half-)parabolas (blue color) are as follows. This solves the first part of the problem.

```
In[·]:= Manipulate[Show[ParametricPlot[
        {x0, t}, {t, 0.1, 8}, PlotStyle ->
        Pink, PlotRange -> {{-10, 10},
        {-2, 8}}], Graphics[{PointSize[
        Large], Yellow, Point[{x0, 0}]}],
        Graphics[{PointSize[Large],
        Yellow, Point[{x0^2, 0}]}],
        ContourPlot[v^2 + 4*u*x0^2 ==
        4*x0^4, {u, -10, 10}, {v, 0.1, 10}],
        ContourPlot[v^2 == 4*y0^2*(u + y0^2),
        {u, -10, 10}, {v, -10, 10}],
        ParametricPlot[{t, y0}, {t, -10,
        10}, PlotStyle -> Pink, PlotRange
        -> {{-10, 10}, {-10, 10}}]],
        {{x0, 0.1}, 1, 10}, {{y0, 0.1},
        1, 10}, SaveDefinitions -> True]
```

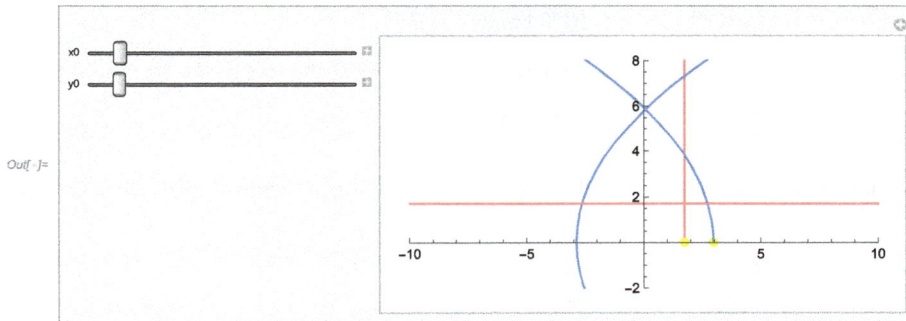

Figure 7.12

The second half of our problem is computing the angle between these two curves at the point $f(z_0)$, where $z_0 = x_0 + iy_0$, $x_0 > 0$, $y_0 > 0$, is the point of intersection of the original lines.

```
In[·]:= {eq1, eq2}
Out[·]:= {v^2 + 4*u*x0^2 == 4*x0^4,
         v^2 == 4*y0^2*(u + y0^2)}
```

We could use techniques from calculus: apply implicit differentiation to find dv/du for both (half-)parabolas and then find the angle between the two tangent lines at the point of the intersection. In this case we also need to use the fact that $\arctan 1/x + \arctan x = \pi/2$. We can also discuss parametrization of plane curves before calculating the angle. It will be the longer route but more suitable for more complicated mappings and also for understanding the relative angular orientation. It also shows the way the pictures were created. We will need some facts about plane curves.

It is convenient to parametrize the image curves (or (half-)parabolas) as follows:

In[·]:= r1[t1_] := {x0^2 - t1^2/(4*x0^2), t1}

In[·]:= r2[t2_] := {t2^2/(4*y0^2) - y0^2, t2}

In[·]:= eq1
Out[·]:= v^2 + 4*u*x0^2 == 4*x0^4

In[·]:= Simplify[eq1 /. {u -> r1[t1][[1]],
        v -> r1[t1][[2]]}]
Out[·]:= True

In[·]:= eq2
Out[·]:= v^2 == 4*y0^2*(u + y0^2)

In[·]:= Simplify[eq2 /. {u -> r2[t1][[1]],
        v -> r2[t1][[2]]}]
Out[·]:= True

The parabolas intersect at a point which can be found by equating the equations in the parametrization. Clearly, $t_1 = t_2 = t$ and

In[·]:= FullSimplify[Solve[x0^2 - t^2/(4*x0^2)
        == t^2/(4*y0^2) - y0^2, t]]
Out[·]:= {{t -> -((2*Sqrt[x0^2 + y0^2])/
        Sqrt[1/x0^2 + 1/y0^2])}, {t ->
        (2*Sqrt[x0^2 + y0^2])/
        Sqrt[1/x0^2 + 1/y0^2]}}

To simplify the expression above, we use the conditions

In[·]:= Assuming[And[x0 > 0, y0 > 0],
        Refine[% // Simplify]]
Out[·]:= {{t -> -2 x0 y0}, {t -> 2 x0 y0}}

Since we take only the upper half of one of the parabolas, only the second value, or t, is needed. We shall denote it by t_0:

In[·]:= t0 = 2 x0 y0
Out[·]:= 2 x0 y0

In[·]:= r1[t0] == r2[t0]
Out[·]:= True

In[·]:= r1[t0]
Out[·]:= {x0^2 - y0^2, 2*x0*y0}

This is precisely the image of the point (x_0, y_0) under the map f:

In[·]:= fuv /. {x -> x0, y -> y0}
Out[·]:= {u, v} == {x0^2 - y0^2, 2*x0*y0}

An angle between curves at some point is the angle between the tangent vectors at this point. The parametric equations of the tangent lines for our case are as follows (here we use the fact that for the curve $r(t)$ the tangent vector at the point is $r'(t_0)$ and use the parametric equations for the tangent line):

In[·]:= tan1[t1_] := Simplify[{x0^2 - y0^2,
      2*x0*y0} + (D[r1[t], t] /. t -> t0)*t1]

In[·]:= tan1[t1]
Out[·]:= {x0^2 - (t1*y0)/x0 - y0^2,
      t1 + 2*x0*y0}

In[·]:= tan2[t2_] := Simplify[{x0^2 - y0^2,
      2*x0*y0} + (D[r2[t], t] /. t -> t0)*t2]

In[·]:= tan2[t2]
Out[·]:= {x0^2 + (t2*x0)/y0 - y0^2,
      t2 + 2*x0*y0}

Let us visualize the original curves (pink color), the parabolas (blue color), and tangent lines (green lines) at the point of intersection for a particular case.

In[·]:= x0 = 1
Out[·]:= 1

In[·]:= y0 = 1
Out[·]:= 1

In[·]:= Show[ParametricPlot[{x0, t}, {t,
0.1, 8}, PlotStyle -> Pink,
PlotRange -> {{-5, 5}, {-1, 8}}],
ParametricPlot[{t, y0}, {t, -5, 5},
PlotStyle -> Pink, PlotRange ->
{{-10, 10}, {-10, 10}}], Graphics
[{PointSize[Large], Yellow, Point[
{x0, 0}]}], Graphics[{PointSize[
Large], Yellow, Point[{x0^2, 0}]}],
Graphics[{PointSize[Large], Pink,
Point[{x0, y0}]}], Graphics[{
PointSize[Large], Blue, Point[
{x0^2 - y0^2, 2*x0*y0}]}],
ContourPlot[v^2 + 4*u*x0^2 ==
4*x0^4, {u, -5, 5}, {v, 0.1, 10}],
ContourPlot[v^2 == 4*y0^2*(u +
y0^2), {u, -5, 5}, {v, -10, 10}],
ParametricPlot[tan1[t1], {t1,
-2^(-1), 1/2}, PlotStyle ->
Green, PlotRange -> {{-10, 10},
{-10, 10}}], ParametricPlot[tan2
[t2], {t2, -3^(-1), 1/2}, PlotStyle
-> Green, PlotRange ->
{{-10, 10}, {-10, 10}}]]

Out[·]=

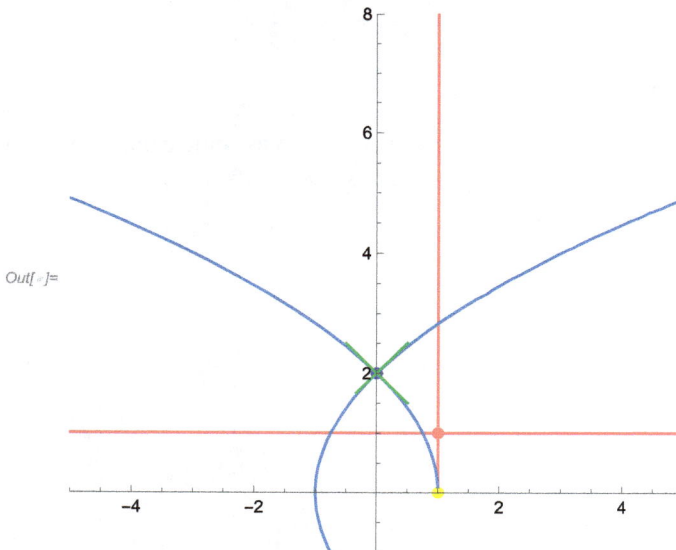

Figure 7.13

In[·]:= Clear[x0]; Clear[y0];

Using the formula for an angle between two tangent vectors, namely

$$\cos\theta = \frac{r_1'(t_0) \cdot r_2'(t_0)}{|r_1'(t_0)||r_2'(t_0)|},$$

for the curves $r_1(t)$ and $r_2(t)$, we easily find the right-hand side

```
In[·]:= Assuming[x0 > 0 && y0 > 0, Refine[
        FullSimplify[D[r1[t], t] . D[r2[t],
        t]/(Norm[D[r1[t], t]]*Norm[D[
        r2[t], t]]) /. t -> t0]]]
```
Out[·]:= 0

So $\cos\theta = 0$. Recall also the directions in which points on the lines moved (directions corresponding to the increasing parameter t). The picture below shows how the tangent vectors to original red and green curves are transformed (and become thicker, larger arrows). Note the relative orientation. This shows that the angle is $\pi/2$.

In[·]:= x0 = 1; y0 = 1;

In[·]:= e1 = 1/4; e2 = 1/4;

```
In[·]:= Show[Show[Graphics[{Red,
        Arrowheads[Large], Arrow[{{x0, y0},
        {x0, y0 + e2}}], Green, Arrowheads[
        Large], Arrow[{{x0, y0}, {x0 +
        e1, y0}}], Red, Thick, Arrowheads[
        Large], Arrow[{({Re[#1^2], Im[#1^2]
        } & )[{x0, y0} . {1, I}], ({Re[
        #1^2], Im[#1^2]} & )[{x0, y0} . {1,
        I}] + (D[r1[t], t] /. t -> t0)}],
        Green, Thick, Arrowheads[Large],
        Arrow[{({Re[#1^2], Im[#1^2]} & )[
        {x0, y0} . {1, I}], ({Re[#1^2],
        Im[#1^2]} & )[{x0, y0} . {1, I}] +
        (D[r2[t], t] /. t -> t0)}]}]]]
```

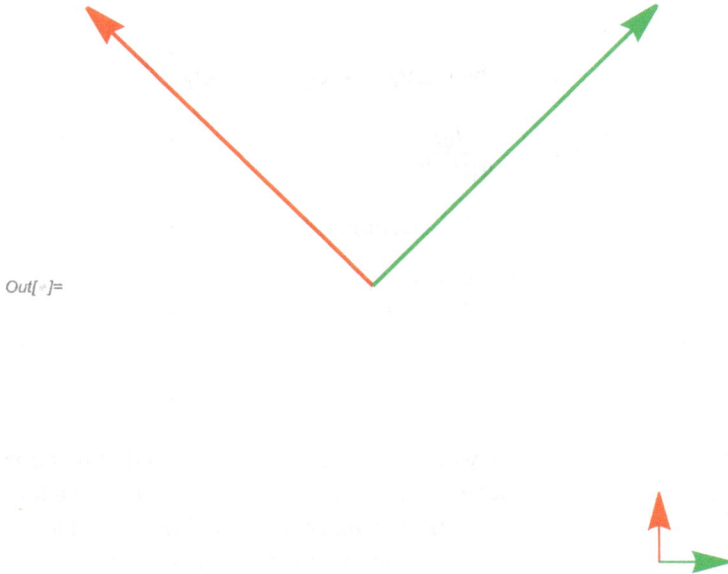

Out[ ]=

Figure 7.14

A conformal transformation is a transformation that preserves an angle between any two intersecting vectors and their relative angular orientation. There are examples when the angle is not preserved. The simplest one is the line through $z = 0$ and $z = 1+i$ and the positive real axis and the mapping $f(z) = \bar{z}^2)$.

7.4 Example 3: The Cauchy residue theorem

In this section we illustrate one of the most famous theorems in complex analysis, namely the Cauchy residue theorem.

Suppose that the function f is analytic in a simply connected domain D except for isolated singularities at $z_1, \ldots, z_n$. Let $\gamma \subset D$ be a closed curve not intersecting any of the singularities and such that all of them are inside. Then

$$\int_\gamma f(z)dz = 2\pi i \left(\sum_{k=1}^{n} \eta(\gamma, z_k)\mathrm{res}(f, z_k) \right).$$

Here $\mathrm{res}(f, z_k)$ denotes the residue of f at the singularity z_k and $\eta(\gamma, z_k)$ is the winding number. We illustrate this below.

The reader can enter the function as a pure function into the input field. The graphic displays the unit circle (in order to provide a sense of scale), the singularities of the function (blue dots), and a piecewise linear path (initially a square) over which integration is performed. One can move the vertices of the path, add new ones (by holding the Control or Alt (Command) key and clicking). The integration is performed numerically so the answer comes out as an approximate number. For an analytic function (without singularities), the integral over a closed curve will always be zero. The imaginary part of the integral over a closed curve which winds around a singularity will be an integer multiple of 2π, though a decimal approximation will usually be displayed.

```
In[·]:= Manipulate[Block[{pp, poles,
        line, z}, pp[ff_] := pp[ff] =
        {ToRules[Reduce[1/ff[z] == 0 &&
        Abs[z - Complex @@ v] < r, z]]};
        poles[ff_] := poles[ff] =
        If[pp[ff] == {}, {}, {Directive[Blue],
        N[Point[{Re[z], Im[z]} /. pp[ff]]]}];
        line = If[cl, Append[pts, First[pts]],
        pts]; Quiet[Column[{Graphics[{
        PointSize[0.05], poles[ff], Pink,
        Arrow[line], Green, Circle[v, r]},
        PlotRange -> All, Axes -> True,
        AxesOrigin -> v, ImageSize ->
        {450, 300}], Text[Style[Text[Row[
        {"\!\(\*SubscriptBox[\(\[Integral]\), \(\[Gamma
        ]\)]\)", ToString[ff[z], TraditionalForm],
        Style["d\[InvisibleSpace]z", Italic, 16], " = ",
        Chop[NIntegrate[ff[z], Evaluate[{z,
        Sequence @@ Apply[Complex, line,
        {1}]}]], 10^(-5)]}]], 20, Black]]}]]],
        {{cl, True, "closed curve"}, {True,
        False}}, {{pts, {{-0.6, -0.6}, {0.6,
        -0.6}, {0.6, 0.6}, {-0.6, 0.6},
        {-0.6, -0.6}}}, v - r, v + r, Locator,
        LocatorAutoCreate -> True}, {{ff,
        #1 & , "function"}, InputField},
        {{v, {0, 0}, "center"}, InputField},
        {{r, 1, "radius"}, InputField},
        SynchronousUpdating -> False,
        SaveDefinitions -> True]
```

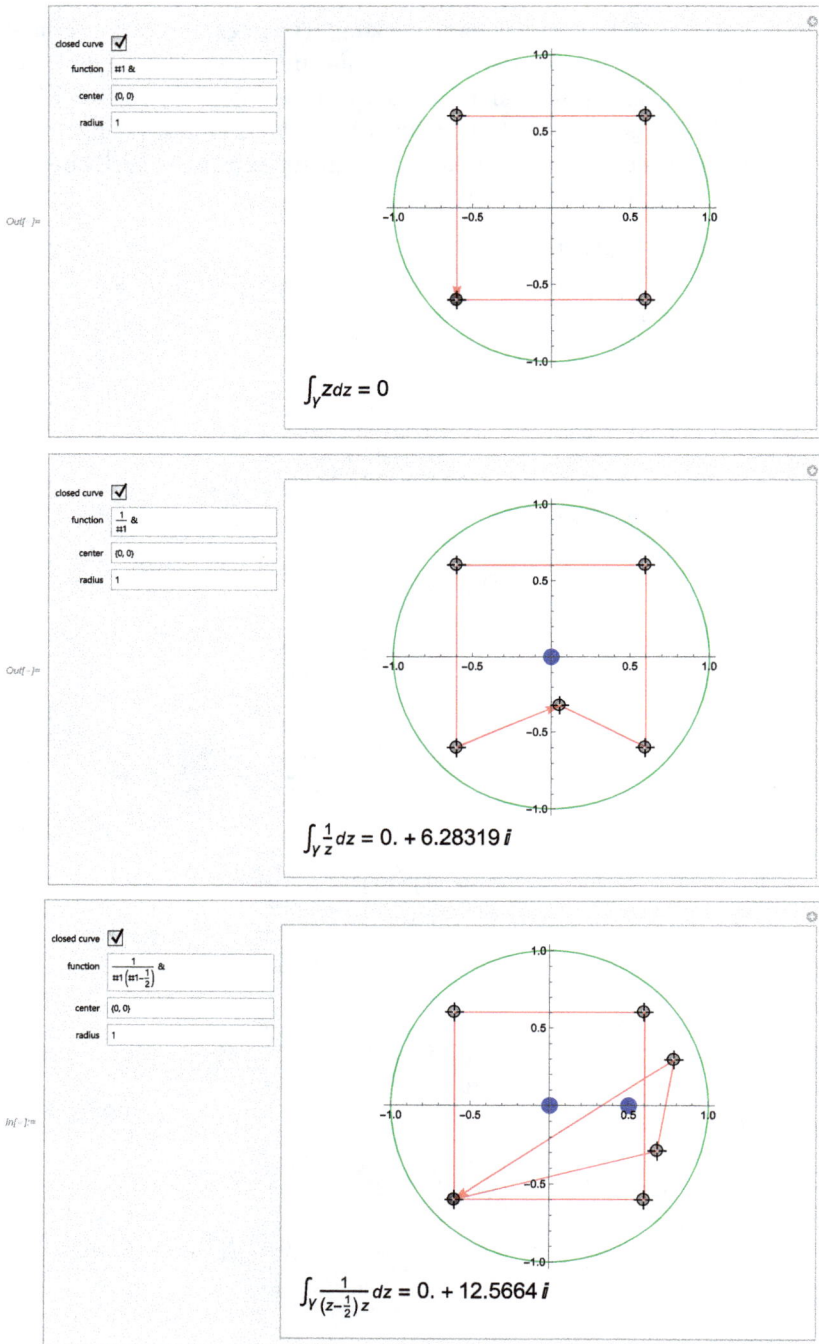

Figure 7.15

Bibliography

[1] Abbena E, Salamon S, Gray A. Modern Differential Geometry of Curves and Surfaces with Mathematica®. Textbooks in Mathematics, Chapman and Hall/CRC, 3rd edition, 2006.
[2] Akhiezer N I. Elements of the Theory of Elliptic Functions. Translations of Mathematical Monographs, Vol. 79, AMS, 1990.
[3] Ahlfors L. Complex Analysis. McGraw-Hill Book Company, 1979.
[4] Chihara T. An Introduction to Orthogonal Polynomials. Dover Publications, Reprint edition, 2011.
[5] Ekedahl T. One Semester of Elliptic Curves. EMS Series of Lectures in Mathematics, European Mathematical Society, 2006 (with Mathematica® notebook available at http://www.math.su.se/~teke/undervisning/Elliptisk.nb).
[6] Filipuk G. The Painlevé equations and orthogonal polynomials. Electronic Notes in Discrete Mathematics 43 (2013), 255–262.
[7] Filipuk G, Kozłowksi A. Analysis with Mathematica®. Volume 1: Single Variable Calculus. De Gruyter Textbook, de Gruyter, 2019.
[8] Filipuk G, Kozłowksi A. Analysis with Mathematica®. Volume 2: Multi-variable Calculus. De Gruyter Textbook, de Gruyter, 2021.
[9] Gautschi W. Orthogonal polynomials, quadrature and approximation: computational methods and software (in Matlab), in: Orthogonal Polynomials and Special Functions. Computation and Applications (F. Marcellan, W. Van Assche, eds.), pp. 1–78. Lecture Notes in Mathematics Vol. 1883, Springer, 2006.
[10] Ishkhanyan T. Mathematical functions and properties, https://www.wolfram.com/broadcast/video.php?v=3273.
[11] Kolmogorov A N, Yushkevich A P (Editors). Mathematics of the 19th Century, Geometry, Analytic Function Theory. Birkhäuser Verlag, Basel–Boston–Berlin, 1996.
[12] Kristensson G. Second Order Differential Equations. Special Functions and Their Classification. Springer, New York Dordrecht Heidelberg London, 2010.
[13] Markushevich A I. The Remarkable Sine Functions. Elsevier, New York, 1966.
[14] Norton R, Abers E. Complex Variables for Scientists and Engineers. Oxford University Press, 2010.
[15] Slavyanov S Yu, Lay W. Special Functions: A Unified Theory Based on Singularities. Oxford Mathematical Monographs, Oxford University Press, 2000.
[16] Shaw W T. Complex Analysis with Mathematica®. Cambridge University Press, 2006.
[17] Van Assche W. Orthogonal Polynomials and Painlevé Equations. Australian Mathematical Society Lecture Series, Vol. 27, Cambridge University Press, Cambridge, 2018.
[18] https://dlmf.nist.gov.
[19] https://en.wikipedia.org/wiki/Envelope_(mathematics).
[20] https://en.wikipedia.org/wiki/Singular_point_of_a_curve.
[21] https://en.wikipedia.org/wiki/List_of_curves.
[22] https://en.wikipedia.org/wiki/List_of_surfaces.
[23] https://en.wikipedia.org/wiki/Tangent_cone.
[24] https://functions.wolfram.com/.
[25] https://mathworld.wolfram.com/.
[26] https://mathworld.wolfram.com/OsculatingPlane.html.
[27] https://mathworld.wolfram.com/SingularPoint.html.
[28] https://reference.wolfram.com/language/tutorial/NDSolveProjection.html.
[29] https://reference.wolfram.com/language/ref/MathematicalFunctionData.html.
[30] https://resources.wolframcloud.com/FunctionRepository/resources/ModularTessellation.
[31] https://resources.wolframcloud.com/FunctionRepository/resources/RiemannSurfacePlot3D.

https://doi.org/10.1515/9783110774641-008

Index

AbsArg 246
AbsArgPlot 38
AbsoluteThickness 10
ArcLength 117, 118, 226
Arrow 8, 170
Assuming 100
Asymptotic 82
AsymptoticDSolveValue 82, 200, 208
AsymptoticIntegrate 82
AsymptoticRSolveValue 82
AsymptoticSolve 82
Axes 15

Ball[n] 58
BoundaryMeshRegion 65, 69
Boxed 15

Canvas 27
Cases 158
Chop 231
Circle 5, 8
Complex 241
ComplexExpand 100
ComplexPlot 36
ComplexPlot3D 224
ComplexRegionPlot 50
Cone 8
Conjugate 241
ContourPlot 4, 107
ContourPlot3D 162, 193
Cuboid 8
Cylinder 8

Dashing 6, 10
DedekindEta 234
DifferenceRoot 204
DifferentialRoot 84
Directives 6, 10
DiscretizeRegion 66
Disk 8, 10
Dot 100
DSolve 73
Dt 230

Eliminate 150
Evaluate 31–33

FrenetSerretSystem 125, 127, 130, 153, 156, 185
FullSimplify 101
FunctionExpand 204
FunctionMeromorphic 237
FunctionSingularities 237

GeometricTransformation 158
Grad 72, 139
Graphics 16, 54, 56
Graphics3D 16, 54, 56, 159
Grid 112
GroebnerBasis 151

HalfLine 70
HalfPlane 70
HalfSpace 70
Hue 10, 179

If 116
ImplicitRegion 56, 59, 109
InfiniteLine 69, 154
InfinitePlane 70
Infinity 158
InputForm 17, 18, 40, 204
Inset 25, 26
InverseJacobiSN 223

JacobiSN 221

KleinInvariantJ 232, 234

LegendreP 204
Line 8, 10
ListPlot 28

Manipulate 15, 43, 76, 116, 187
Maximize 181
MaxValue 182
Mesh 15
MeshRegion 65, 66
MinValue 182
ModularLambda 234

NDSolve 73, 74, 76, 79, 146
NestList 136
Norm 99, 100
Normal 18

https://doi.org/10.1515/9783110774641-009

Normalize 100, 170

Opacity 10

ParametricNDSolve 74, 76, 77
ParametricNDSolveValue 80
ParametricPlot 3, 16, 101, 111
ParametricPlot3D 3, 16, 19, 101, 111, 158, 161, 162
ParametricRegion 58
ParamtericPlot3D 158
PlaneCurve 89
PlaneCurveData 89
Plot 1, 16, 17
Plot3D 1, 16
PlotRange 1
Point 8, 10
PointSize 10
Polygon 8
Projection method 79

Raster 8
Rectangle 8
Red 10
Reduce 70, 101, 226
Region 51, 53, 54, 56, 65, 66, 94, 104
RegionMeasure 103, 118
RegionPlot 16, 54
RegionPlot3D 16, 54
Regions 113
Relm 39, 244
RelmPlot 35
Rescale 182
Residue 198

Resultant 150
RGBColor 10, 182
RotationTransform 127, 158, 162

Series 198
SetAttributes 39
Short 17
Show 6, 13, 21, 23, 112
Simplex 8
Simplify 100
SpaceCurve 89, 92, 152
SpaceCurveData 89, 96
Sphere 8
Sphere[n] 58
StreamPlot 139
StreamPlot3D 140
Surface 89, 92
SurfaceData 89

Table 149, 181
Text 8
Thick 10
Thickness 10
Thin 10
Through 190
TransfomationFunction 158
TranslationTransform 158
Transpose 170
Tube 8, 162, 170

VectorQ 99

WeierstrassP 232
WorkingPrecision 225

www.ingramcontent.com/pod-product-compliance
Lightning Source LLC
Chambersburg PA
CBHW061353210326
41598CB00035B/5967